MICROBIOLOGY FOR THE HEALTH PROFESSIONS

LABORATORY MANUAL

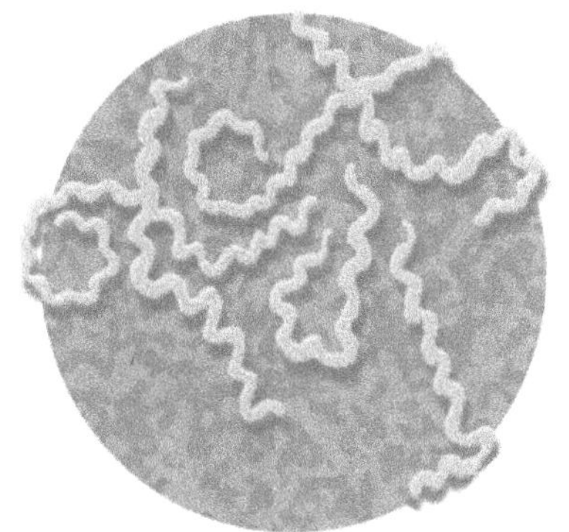

Spirillum
(corkscrew-shaped)

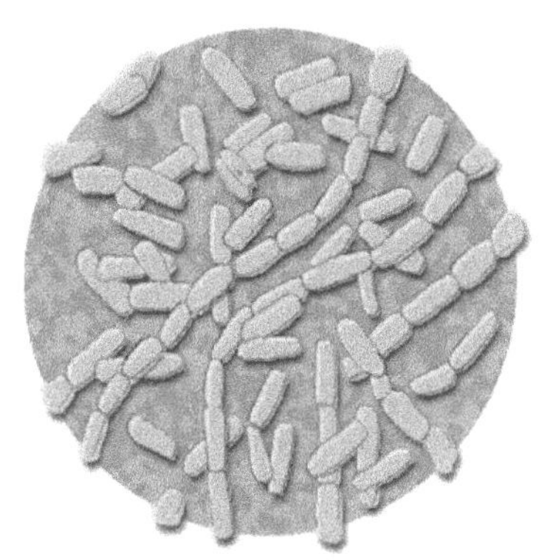

Bacillus
(rod-shaped)

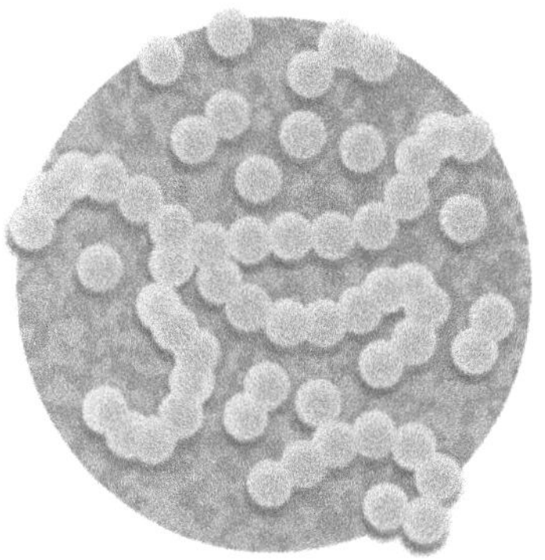

Coccus
(spherical)

Diana Fagan, Ph.D.
Jonathan Caguiat, Ph.D.

Cover © Shutterstock.com
Interior illustrations created by Jonathan Caguiat unless stated otherwise

Kendall Hunt
publishing company

www.kendallhunt.com
Send all inquiries to:
4050 Westmark Drive
Dubuque, IA 52004-1840

ISBN: 978-1-5249-8621-6

Published in the United States of America

Dedication

Dedicated to Dr. Paul D. Van Zandt, the originator of Microbiology for the Health Professions, Biological Sciences Faculty for 30 years and Chair for 21 years.

Contributing Authors

Leonard A. Perry, MPH and Sandra Senedak, B.S.
Food Microbiology Manual. Published by Youngstown State University.

Leonard A. Perry, MPH, Sandra Senedak, B.S., and Paul Van Zandt, Ph.D.
Microbes in Focus: A Laboratory Manual of Microbiology for the Allied Health Professions. Published by Youngstown State University.

CONTENTS

ACKNOWLEDGMENTS

This work was supported by the Department of Biological Sciences, in the College of Science, Technology, Engineering and Math, and by the Office of Environmental and Occupational Health and Safety at Youngstown State University, Youngstown, OH.

INTRODUCTION

The primary purpose of this manual is to acquaint nursing and other allied health professions students with potential problems associated with microbial contamination that may occur in hospital and clinical environments. The major focus of the experiments presented in this manual is the bacteria. However, students should realize that other types of microbes, such as viruses and fungi, also contribute to nosocomial (hospital acquired) morbidity and mortality. The carefully designed experiments in this manual show how microbial infections may occur and what methods reduce their appearance. It is not the intention of this manual to teach the student all the techniques involved in microbial detection and identification. This manual only introduces the basic techniques necessary to perform certain experiments.

All students must learn aseptic and sterile techniques to complete experiments successfully. Once developed, they should never forget these techniques because most clinical environments require aseptic conditions.

Many experiments will use pure cultures. A pure culture contains only one species of microorganism, that is, it is free from all other types of microorganisms. If a student works carefully and follows the directions, a pure culture will remain pure indefinitely. A careless student will contaminate a pure culture instantly.

This laboratory course uses microorganisms that are considered nonpathogenic, that is, they usually do not cause serious diseases. However, any microbe might cause an infection under certain conditions when introduced into the body accidentally. A student with a compromised immune system must be especially aware that the microbes used in this laboratory course can cause opportunistic infections. Therefore, it is good laboratory practice to treat all bacteria as potential pathogens. It is for this reason that the following rules must be followed to ensure that the laboratory is a safe place to work.

1. Disinfect hands and laboratory benches at the beginning of each laboratory period.
2. Wear a laboratory coat and goggles. Your coat or jacket should be made of material that can be sterilized. After each laboratory session, it is good practice to store your coat or jacket in a plastic bag. At the end of the course, the coat or jacket can be sanitized by washing it in hot water and/or bleach.
3. Avoid spilling material. If infectious material comes in contact with the laboratory bench, the floor, hands, or clothing, notify the instructor at once.
4. All cultures or infectious material must be disposed of in such a manner as to be of no danger. Special containers are provided for this purpose.
5. Glassware and other equipment should be kept clean and in their proper place.
6. Fingers often become contaminated. Do not touch your face with your hands. Cover cuts and scrapes with band aids.
7. Do not eat or drink in the laboratory. Do not lick gummed labels or put pencils in your mouth.
8. Arrange cultures in a secure place on the laboratory bench. Avoid accumulation of unused equipment on the laboratory bench.
9. Always sterilize inoculating loops thoroughly before laying them down.
10. Be sure that you take proper care of your microscope (see Laboratory Exercise 1).
11. Disinfect hands and laboratory benches before leaving the laboratory.
12. Points will be deducted from the final laboratory score if rules are violated.

LABORATORY EXERCISE 1

THE USE AND CARE OF THE MICROSCOPE: VIEWING PREPARED SLIDES OF MICROORGANISMS

Materials Needed:

Prepared microscope slides of bacillus, coccus, and spirillum-shaped bacterial cells.

Purpose:

To show students how to use a microscope properly and introduce commonly observed bacterial shapes.

Background Information:

Figure 1–1A is a picture of a bright-field microscope. Acquainting yourself with the names of the common parts will make this exercise easier to follow. The condenser focuses light from the light source on the specimen (Fig. 1–1B). The light that is not deflected by the specimen passes through the objective and ocular lenses, producing a colored image of the specimen against a bright background.

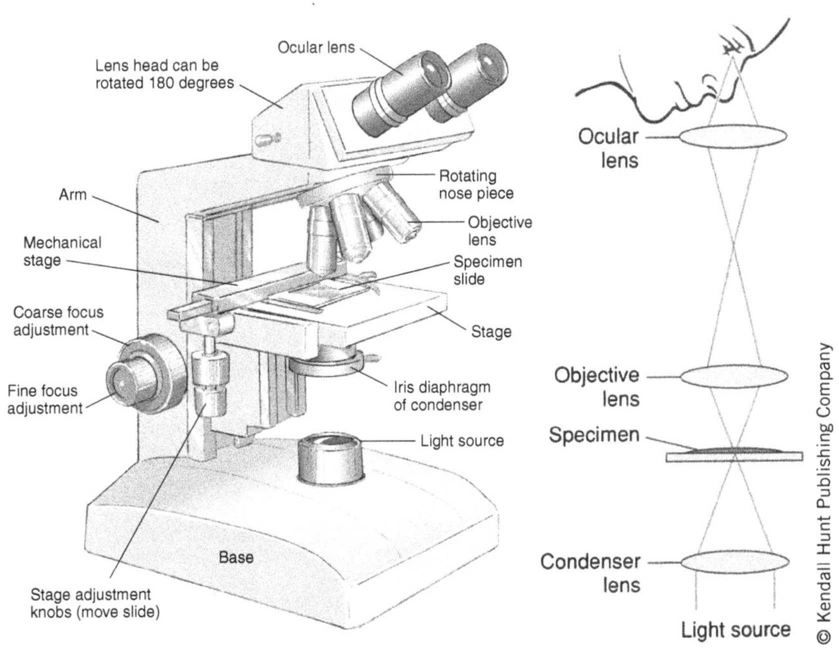

Figure 1–1. The bright-field microscope. (A) Parts of a bright-field microscope. (B) Light path in a bright-field microscope.

Both lenses magnify the image. The ocular lens or eyepiece magnifies the image by 10×, and the objective lens magnifies it by 4×, 10×, 40×, or 100×, depending on the lens that is being used. When the low-power (10×) objective lens is used, the total magnification is 10 × 10 = 100×, because the light passes through both lenses. Table 1–1 lists the magnification properties of each objective. The scanning (4×) and low-power (10×) objectives are used to locate a specimen on the slide, and the high-dry (40×) and oil immersion (100×) objectives are used to view fine details of a specimen.

Table 1–1. Magnification for each objective lens.

Objective	Magnification of Objective	Eyepiece Magnification	Total Magnification
Scanning	4×	10×	40×
Low power	10×	10×	100×
High dry	40×	10×	400×
Oil immersion	100×	10×	1,000×

When light passes through a slide or a lens, it is refracted. Refraction is the process in which light bends when it passes from a material of one density (glass) to a material of a different density (air). When light is refracted as it passes through a specimen to the air, the low-power (10×) and high-dry (40×) lenses still capture it. However, the 100× objective is smaller and fails to capture the light, because the bent light falls outside the lens (Fig. 1–2). Immersion oil has the same refractive index as the glass in the slide holding the specimen and the glass in the lens, preventing refraction and improving resolution (the ability to see the sample in focus). The oil immersion (100×) objective is required to view the morphologic characteristics of microorganisms because they cannot be seen at a magnification below 1,000×.

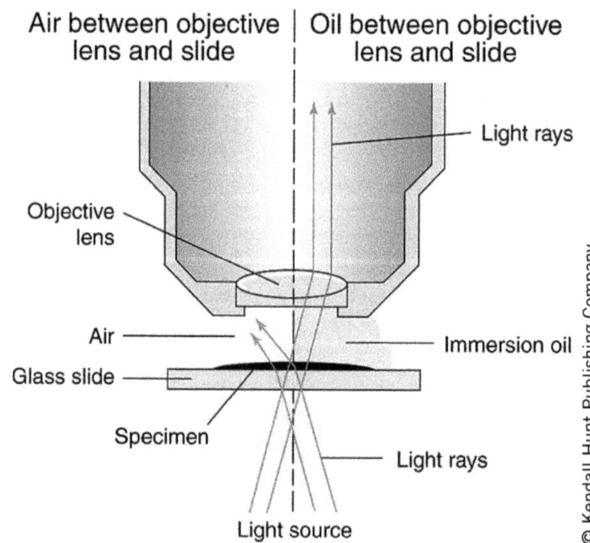

Figure 1–2. Importance of using immersion oil for the 100× objective. Left: In the absence of immersion oil, refraction prevents the light from entering the lens. Right: Immersion oil has the same refractive index as the slide and lens, preventing refraction and allowing the light to enter the lens.

Care of the Microscope. You are expected to observe the following instructions regarding the proper care of the microscope that you are assigned this semester.

1. You are responsible for the microscope you use. Keep the stage of your microscope clean. Clean the lenses with lens paper only. Never use anything (paper towels, Kim Wipes, etc.) other than lens paper to clean the microscope lenses. Never remove the lens or ocular piece.

2. When using the oil immersion lens (100×), always use a drop of the special immersion oil.
3. Use two hands to carry or handle your microscope; always handle it gently.
4. Use the coarse focus knob only with the scanning (4×) and low-power (10×) objectives because they cannot come in close contact with the slide on the stage at any level.
5. Use the fine focus knob with the 40× and 100× objectives only after focusing first with the 10× objective because they can come in close contact with the slide on the stage. Hitting the slide with an objective may damage the lens.
6. When you have finished with your microscope, clean all lenses and ocular pieces with lens paper to remove excess dust and oil. Place the low-power objective into place and put the microscope in its proper place.
7. If there is any malfunction with your microscope, do not try to repair it yourself. Report the problem at once to your laboratory instructor.

How to Focus the Microscope. The following instructions provide a step-by-step procedure for using your microscope to view microbes. For this procedure, you will be given prepared slides. Take all necessary precautions to protect this slide and return it when finished.

Focusing:

1. Place the microscope directly in front of you on the laboratory bench and plug it in. Turn the light switch on using the switch at the front of the microscope. The light intensity can be adjusted for your comfort using the rheostat on the side of the base.
2. Adjust the condenser so that it is slightly below the stage to allow the maximum amount of light. If it is too high, you will produce a light artifact that looks like pink and blue grains of sand. If it is too low, the sample will appear dark. Make sure that the iris diaphragm is completely open so that you can see light coming through the condenser.
3. Place the slide on the stage and secure it between the clips. Then, mechanically center your slide so that the center of it is directly over the lens of the condenser.
4. Focus the microscope on the specimen using the low-power (10×) objective first. Start with the objective as close to the slide as possible. While looking through the ocular lens (eye piece), move the stage down using the course adjustment knob until the specimen is focused. It may be necessary to move the slide with the stage adjustment knobs to find the specimen.
5. After focusing at low power (10×), swing the high-dry (40×) objective into place. The specimen should still be in focus. Sharpen the focus only with the fine adjustment knob. The microscope is parfocal, meaning that when a microscope is focused using one objective, it will stay in focus when switched to another objective.
6. To use the oil immersion lens, position the 4X objective pointing down (do this without changing the position of the stage so that the sample remains in focus). Place a very small drop of oil on the slide. Swing the oil immersion (100×) objective into the oil while watching from the side. Again, since the microscope is parfocal, the specimen should be in focus. Sharpen the focus only with the fine adjustment knob.
7. If you lose focus on the specimen, go back to step 4 to refocus the microscope. If you are continuously unsuccessful in focusing, ask your course instructor or teaching assistant for help.
8. Observe the slides with careful attention. Make drawings of what you see. The drawings should only include four to five individual organisms and should show the clustering of the organisms. Make these drawings large enough so that you can tell the shape of the organism. Indicate the total magnification used.
9. When finished for the day, raise the objective from the slide and gently wipe the oil from the lens and slide. Return your microscope to its proper place, clean and in good condition.

Remember to use only lens paper to clean oil from the microscope objective. Use Kim Wipes or a similar type of tissue to clean the slides and the surface of the microscope.

2. When using the oil immersion lens (100×), always use a drop of the special immersion oil.
3. Use two hands to carry or handle your microscope. Always handle it gently.
4. Use the coarse focus knob only with the scanning (4×) and low power (10×) objectives, because they cannot come in close contact with the slide on the stage at any level.
5. Use the fine focus knob with the 40× and 100× objectives only, after focusing first with the 10× objective, because they will come in close contact with the slide on the stage. Hitting the slide with an objective may damage the objective.
6. When you have finished with your microscope, clean all lenses and ocular pieces with lens paper to remove excess oil and dust. Place the low-power objective into place and put the microscope in its proper place.
7. If there is any malfunction with your microscope, do not try to repair it yourself. Report the problem at once to your laboratory instructor.

[illegible]

LABORATORY EXERCISE

2 INOCULATION OF AGAR PLATES WITH BACTERIA FROM THE ENVIRONMENT

Materials Needed:

Nutrient Agar (NA) plates (5)
Cotton swabs (sterile)
Sterile water

Purpose:

To show that microbes are everywhere, thereby emphasizing the need for aseptic technique.

Background Information:

Bacteria and other microorganisms exist any place where water and nutrients are available for their survival. The bacteria in our environment can include pathogens, but most are harmless or actually provide benefit to us. Bacteria and other microorganisms (protozoa and fungi) are important for decomposing dead organisms and returning nutrients to the soil. Bacteria are also found in large numbers on most surfaces of the body, due to the abundance of nutrients. Most of these bacteria are normal flora, organisms found on the body without causing harm. In addition, some provide needed nutrients and vitamins and prevent the growth of pathogens. In this exercise, you will sample your environment for the presence of bacteria. Any bacterium placed on a nutrient agar plate will multiply rapidly to produce a visible colony. In this way, you can detect the presence of microorganisms in your environment.

Procedure:

1. For one of the nutrient agar plates, remove the lid and place the bottom half containing the medium at some location within the laboratory. Expose the surface of the agar to the air for 10 to 15 minutes before replacing the lid.
2. The remaining plates may be contaminated in any of the following ways:
   - coughing on the plate
   - rubbing fingers across the plate
   - swabbing a desk, phone, or other object and then swabbing the surface of the plate
3. Label the source of contamination on the bottom (agar side) of each plate. (NOTE: Always label the bottom of your plates, not the lid.)
4. Place the plates at room temperature for 1 week in an inverted position with the lid down and the medium side up.
5. Examine the plates for microbial growth during the next laboratory period. Each bacterium divides thousands of times to produce a visible colony, which will have a unique appearance for each type of bacterium. Record your results using drawings and verbal descriptions.

LABORATORY EXERCISE

3 MICROSCOPIC OBSERVATION OF BACTERIA USING THE SIMPLE STAIN PROCEDURE

Materials Needed:

Nutrient Agar (NA) plate of *Bacillus subtilis*
Nutrient Agar (NA) plate of *Escherichia coli*
Nutrient Agar (NA) plate of *Staphylococcus epidermidis*
Methylene blue stain
Glass slides
Coverslips
Toothpicks or cotton swabs

Purpose:

To learn how to view bacteria under a microscope using a simple stain and observe their shapes. Also, to view eukaryotic cells under the microscope and see that they are larger than bacterial cells.

Background Information:

Simple staining makes it possible to see the size and shape of colorless bacteria. It uses only one dye, methylene blue, which is used to stain a wide variety of microorganisms. It is a basic dye, which is positively charged, allowing it to bind to negatively charged structures in cells.

Procedure:

Students will use an incinerator to sterilize inoculating loops. Before using the incinerator, turn it on until the inside becomes red hot. Place the loop in the incinerator until the metal wire turns red. Always hold the loop while it is in the incinerator. Never leave the loop hanging in the incinerator. This will melt the handle and damage the loop. It may also make the handle so hot that it will burn the hand of the next person who removes the loop. Once the metal wire turns red, remove the loop and allow it to cool in the air. Do not lay it on the table and do not touch it to see if it is cool. If it sizzles when you touch bacteria with it, then it is killing the bacteria. Use an incinerator to sterilize the loop between steps. A Bunsen burner may also be used to sterilize a loop. However, heating a loopful of bacteria in an open flame can create aerosols, causing some bacteria to become airborne. Incinerators prevent this because the heat source is in an enclosed space. Bacteria in an aerosol produced by an incinerator cannot escape and will be killed inside the hot enclosed space.

A. *Staining a Smear Prepared from an Agar Culture.*
   1. Transfer a small amount of water to a glass slide using an inoculating loop.
   2. With a wax pencil, draw lines on each side of the drop of water.

3. Sterilize the inoculating loop in the incinerator and hold it in the air for a few seconds to allow it to cool.
4. Transfer a small amount of bacteria from an agar plate to the water drop on the slide. Mix the bacteria and water to form a smooth suspension between the wax pencil lines.
5. Allow the slide to air dry. Use a heating block set on low to speed drying if there is too much water.
6. Heat fix the bacteria to the slide by passing the smear quickly through a Bunsen burner flame three times. Do not hold the slide in the flame.
7. Completely cover the smear with methylene blue and allow the stain to react for 1 minute.
8. Holding the slide by its edges over the sink at a 45° angle, decolorize the smear by dripping water down the slide until the water that is dripping off the slide runs clear.
9. Blot the slide dry with a paper towel.
10. Examine the smear under the oil immersion objective of a microscope and record your results using drawings and verbal descriptions.
11. Repeat steps 1 to 10 for the other bacterial strains.

B. *Staining Material Collected from the Mouth Using a Wet Mount.*

1. Using a sterile inoculating loop, place a drop of water on a clean glass slide.
2. Scrape off plaque by the edge of your gum using a toothpick or the handle of a cotton swab.
3. Mix the debris from your mouth into the water droplet on the slide. Do not allow it to dry.
4. Place a glass coverslip over the preparation (Fig. 3–1) and use the low-power (10×) objective of a microscope to focus on the edge of the coverslip.

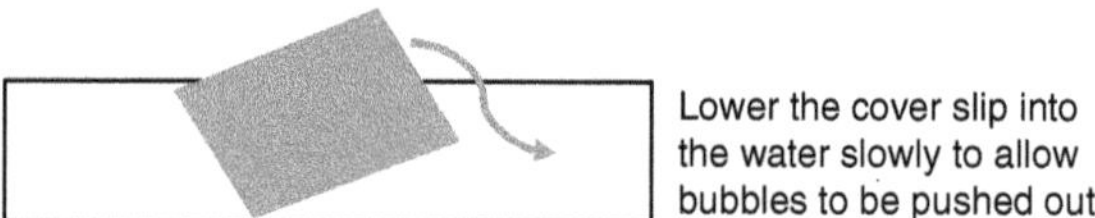

Figure 3–1. Placing a coverslip on a glass slide.

5. Switch to the high-dry (40×) objective and refine the focus on the edge of the coverslip using the fine adjustment knob. Do not use oil.
6. Add one drop of methylene blue to the edge of the coverslip so that the stain diffuses under the coverslip (Fig. 3–2).

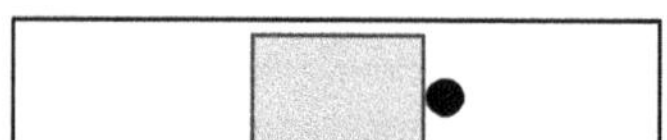

Figure 3–2. Place a drop of methylene blue on the edge of the coverslip.

7. View the specimen using the high-dry (40×) objective at the edge of the dye front. Be extremely careful to not let the stain touch the objectives or spill onto the stage or condenser of the microscope.
8. Record your observations using drawings and verbal descriptions. Look for epithelial cells, white blood cells, and plaque.

OBTAINING ISOLATED COLONIES BY THE POUR PLATE METHOD

Materials Needed:

Nutrient Broth (NB) culture containing a mixture of *Staphylococcus epidermidis*, *Micrococcus luteus*, and *Klebsiella pneumonia*
Nutrient Agar (NA) molten deep (held at 45°C–50°C in a heat block/water bath)
Petri dish (sterile)

Purpose:

To separate different types of bacteria from a mixture of bacteria.

Background Information:

This method may be used to determine the number of viable bacteria in a culture. Since one isolated microbe gives rise to one bacterial colony, counting the number of colonies determines the number of bacterial cells that were added to the plate.

Procedure:

1. Allow a tube of molten nutrient agar to cool until you can hold it.
2. Using a sterile inoculating needle, transfer a sample of the mixed culture to the molten agar (Fig. 4–1). Be sure that the bacteria in the mixed culture has been fully resuspended prior to taking the sample.

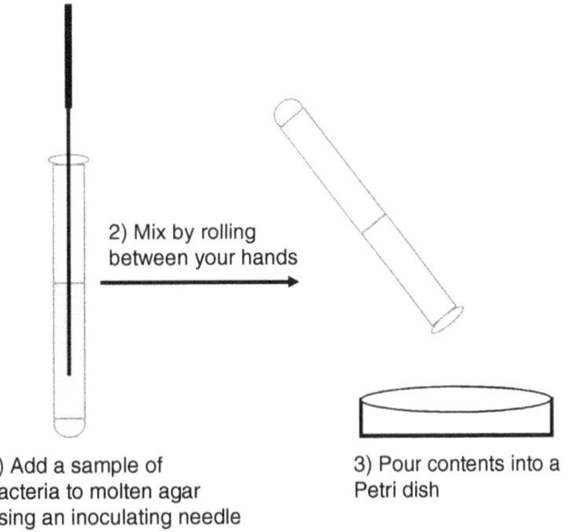

Figure 4–1. Diagram for steps 2 to 4 for the pour plate exercise.

3. Roll the tube quickly between your hands to mix the bacteria into the agar.
4. Pour the contents of the tube into a Petri dish.
5. Swirl the dish gently to ensure that the agar completely covers the bottom of the plate.
6. Allow the agar to solidify at room temperature for approximately 10 minutes.
7. Label the plate on the bottom (agar side) and place it upside down with the agar side up at room temperature for 1 week.
8. Record the results using drawings and verbal descriptions.

LABORATORY EXERCISE

5 OBTAINING A PURE CULTURE BY THE STREAK PLATE METHOD

Materials Needed:

Nutrient Broth (NB) culture containing a mixture of *Staphylococcus epidermidis*, *Micrococcus luteus*, and *Klebsiella pneumonia*
Nutrient Agar (NA) plate

Purpose:

To separate different types of bacteria from a mixture of bacteria.

Background Information:

This method also assumes that one isolated microbe gives rise to one bacterial colony. The cross-hatched streaking technique distributes the bacteria across the plate to produce well-defined bacterial colonies. If a number of different organisms are present in a specimen, the streak plate method provides a means by which each type of organism may be isolated for further studies. It cannot be used to determine the number of viable bacteria that are present in a culture.

Procedure:

1. Draw four quadrants with a wax pencil on the bottom of a nutrient agar plate, as shown in Figure 5–1A.
2. Sterilize a loop, let it cool, obtain a loopful of bacteria and streak in quadrant 1, as shown in Figure 5–1B. Do not dig the loop into the agar. Gently slide it across the surface of the agar. Do not pick up more bacteria for steps 3 and 4. The goal is to spread fewer and fewer bacteria in each quadrant so that single bacteria can grow to form isolated colonies.
3. Sterilize the loop, let it cool, drag it through quadrant 1 twice, and continue streaking in quadrant 2.
4. Repeat for quadrants 3 and 4 as shown in Figure 5–1, remembering to sterilize the loop before streaking each quadrant.
5. Incubate the plate upside down with the agar side up at room temperature for 1 week.
6. Record your results using drawings and verbal descriptions.

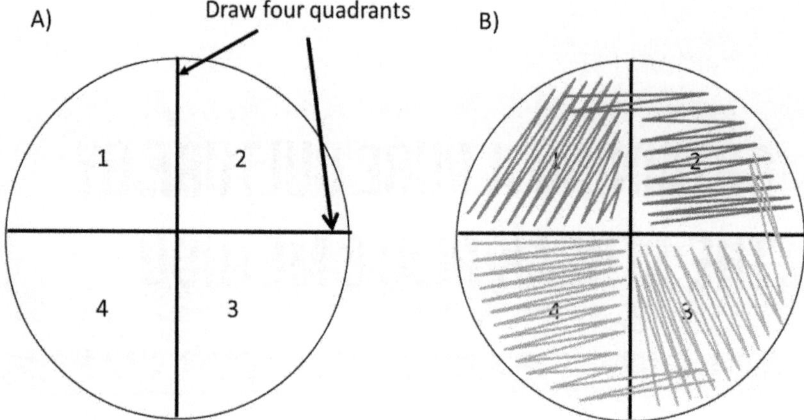

Figure 5–1. The streak plate technique. (A) Plate with four quadrants. (B) Streak plate pattern. Spread bacteria with a loop in quadrant 1 first. Then, spread bacteria from quadrant 1 into quadrant 2, and from quadrant 2 into quadrant 3 and from quadrant 3 into quadrant 4. Sterilize the loop before streaking from one quadrant to the other.

LABORATORY EXERCISE

6 DIFFERENTIAL STAINING OF BACTERIA: THE GRAM STAIN

Materials Needed:

Nutrient Agar (NA) culture of *Staphylococcus epidermidis*
Nutrient Agar (NA) culture of *Escherichia coli*
Nutrient Agar (NA) culture of *Bacillus subtilis*
Glass slides

Purpose:

To show the method of differentiating bacteria by Gram-staining.

Background Information:

The Gram stain is one of the most frequently used stains in the microbiology laboratory (Snyder, 1970). It is a differential stain because it distinguishes between different microorganisms. A smear of the test microorganism is stained with a solution of crystal violet. Then, a solution of iodine, which acts as a mordant, fixes the stain to the cell. When the smear is decolorized with acetone, the thick peptidoglycan layer of gram-positive bacteria retains the dye, but the thin layer of peptidoglycan in gram-negative bacteria does not retain it. Counterstaining the smear with safranin makes it possible to see gram-negative bacteria. Gram-positive bacteria appear blue or purple, whereas gram-negative bacteria appear red or pink under the oil immersion objective of a microscope. Gram-staining is an important technique for microbe identification. When performing the Gram stain, it is important to use a culture that is not more than 24 hours old. Older cultures exhibit variable Gram-staining characteristics.

Purpose:

NOTE: This procedure describes the Philadelphia General Hospital (PGH) method. Other Gram-staining methods are similar and give the same results.

1. Prepare separate smears on glass slides for *S. epidermidis*, *B. subtilis*, and *E. coli* as described in Laboratory Exercise 3.
2. Flood each smear with crystal violet.
3. Add three to eight drops of the sodium bicarbonate solution to the slide. Rock the slide gently for 5 to 10 seconds. Pour off excess liquid into the sink by holding the slide at a 45° angle.
4. Flood the smear with iodine solution. Rock the slide gently for 5 to 10 seconds.

5. Holding the slide at a 45° angle, wash the smear with acetone (decolorizer) until it runs colorless from the slide. Let the slide dry completely for a few seconds or rinse it with water.
6. Flood the smear with the safranin solution. Rock the slide gently for 1 to 2 minutes.
7. Rinse the slide with water and gently blot it dry.
8. View the smear under the oil immersion objective of a microscope.
9. Record the results using drawings and verbal descriptions.

LABORATORY EXERCISE

7

HYDROLYSIS OF STARCH

Materials Needed:

Nutrient Agar (NA) culture of *Escherichia coli*
Nutrient Agar (NA) culture of *Bacillus subtilis*
Starch agar plates (2)

Purpose:

To test for starch hydrolysis in two different bacterial strains.

Background Information:

Bacteria, like all living organisms, break down complex chemical substances into simple ones that they import and use for metabolism. In order to accomplish this, bacteria produce highly specific enzymes. An enzyme is an organic catalyst that speeds up the rate of a chemical reaction. An enzyme acts on a substance called a substrate, and each enzyme recognizes a specific substrate. This recognition results in a specific reaction between an enzyme and its substrate.

Starch agar is a nutrient medium containing a small amount of starch. One of the microorganisms that will be streaked on the agar surface produces amylase, an enzyme which hydrolyzes starch. This enzyme is secreted outside the cell because starch is too large to import into the cell. Amylase breaks it down into smaller molecules of maltose and glucose, which are imported into the cell and used as a carbon source. Iodine reacts with starch to form a brown or blue color. When flooded with iodine, bacterial streaks that secrete amylase will produce a clear halo, whereas bacterial streaks that do not produce amylase will not produce a clear halo.

Procedure:

1. Obtain two starch agar plates. Label the bottom of one with *E. coli* and label the other with *B. subtilis.*
2. Using aseptic technique and your inoculating loop, make a single streak down the middle of each plate with the appropriate organism, as shown in Figure 7–1. Do not go to the edges of the plate with the streak. There needs to be enough agar surrounding the bacteria so that you can see a halo, if amylase is produced.

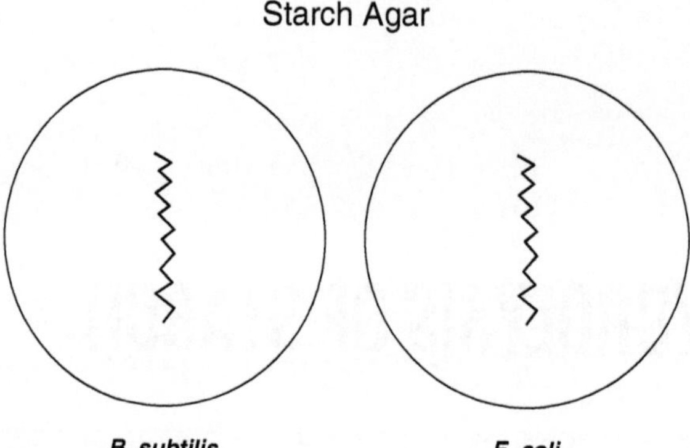

Figure 7–1. Streak each strain down the center of a starch agar plate.

3. Incubate the plates at room temperature for 1 week.
4. Remove the lid of each plate, then flood the media surfaces with an iodine solution.
5. Record the results using drawings and verbal descriptions.

LABORATORY EXERCISE

HYDROLYSIS OF GELATIN

Materials Needed:

Nutrient Agar (NA) culture of *Proteus mirabilis*
Nutrient Agar (NA) culture of *Bacillus subtilis*
Nutrient Agar (NA) culture of *Escherichia coli*
Nutrient gelatin tubes (3)

Purpose:

To test for gelatinase activity in three different bacterial strains.

Background Information:

This exercise examines the ability of three different bacterial strains to produce the enzyme, gelatinase. Each strain will be introduced into a medium containing gelatin. Gelatin forms a solid support in the tube when it is cold. If the bacterium digests it, the medium in the tube will remain liquid when it is refrigerated. If the bacterium cannot digest it, the medium in the tube will solidify when it is refrigerated.

Procedure:

1. Inoculate three tubes of nutrient agar with one of the three strains of bacteria using a needle (Fig. 8–1). The nutrient gelatin medium is a liquid at room temperature.
2. Incubate the tubes at 37°C for at least 5 days to allow the bacteria to grow.
3. Remove tubes from the incubator and place them in a 5°C to 10°C refrigerator for about 20 minutes. Determine whether the gelatin has solidified or has remained liquid.
4. Record the results using drawings and verbal descriptions.

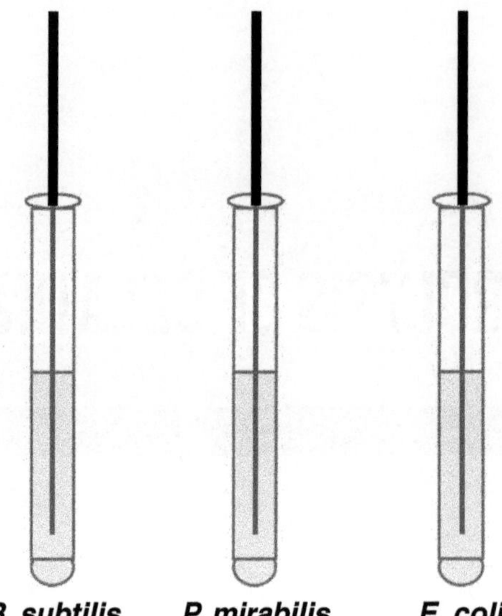

Figure 8–1. Three tubes of nutrient gelatin medium are inoculated with one of the bacterial strains.

LABORATORY EXERCISE

TYPES OF MICROBIOLOGICAL MEDIA

Materials Needed:

18- to 48-hour MacConkey (MAC) agar plate inoculated with *Escherichia coli*
18- to 48-hour MacConkey (MAC) agar plate inoculated with *Alcaligenes faecalis*
18- to 48-hour Eosin Methylene Blue (EMB) agar plate inoculated with *Escherichia coli*
18- to 48-hour Eosin Methylene Blue (EMB) agar plate inoculated with *Alcaligenes faecalis*
18- to 48-hour Mannitol Salt Agar (MSA) plate inoculated with *Staphylococcus epidermidis*
18- to 48-hour Mannitol Salt Agar (MSA) plate inoculated with *Staphylococcus aureus*

Purpose:

To learn about different types of microbiological growth media.

Background Information:

In order to study bacteria in the laboratory, they must first be grown or cultured on or in a growth medium. A microbiological medium may be in liquid or semi-solid (gel) form. A liquid form is referred to as a broth medium, whereas a semi-solid form is referred to as an agar medium. Agar, a derivative of seaweed, is usually used as the solidifying agent and is not hydrolyzed by microorganisms. Hence, when mixed with growth medium, autoclaved, and cooled, the agar solidifies to provide a semi-solid medium that will support the growth of many types of microorganisms. Agar is a much more suitable solidifying agent than gelatin, since gelatin liquefies above room temperature and is easily hydrolyzed by many microorganisms.

Microbiological media fall into one of three categories: nutrient, differential, and selective (www.sharlab.com).

- A nutrient medium is any medium that will support and maintain the growth of a wide variety of microorganisms.
- A differential medium sustains the growth of a large number of bacteria and distinguishes one type of bacterium from another.
- A selective medium allows one type of bacterium to grow but prevents the growth of others.

Nutrient broth and nutrient agar are examples of nutrient media. Some microbes are fastidious and require a nutrient media enriched with additional specific requirements. Blood agar is an example of an enrichment medium. It is supplemented with 5% blood from sheep or other animals. Since blood makes it highly enriched, it is suitable for growing many fastidious microorganisms, including human pathogens.

MacConkey (MAC) agar is a selective and differential medium. Bile in this medium inhibits the growth of Gram-positive bacteria but allows Gram-negative bacteria to grow. This medium also differentiates between

Gram-negative bacilli that ferment lactose and those that do not. Bacteria that ferment lactose appear as brick red colonies, while those that do not ferment lactose appear as colorless colonies.

Mannitol Salt Agar (MSA) is a selective and differential medium. A high level of salt (sodium chloride) inhibits the growth of salt-intolerant organisms, allowing *Staphylococcus* bacteria to grow. It also differentiates between organisms that ferment the sugar, mannitol. Bacteria that ferment mannitol cause the surrounding medium to turn yellow, indicating the presence of *Staphylococcus aureus*. Bacteria that do not ferment lactose do not cause a color change.

Eosin Methylene Blue (EMB) agar is a selective and differential medium. The dyes, eosin and methylene blue, inhibit the growth of gram-positive bacteria and allow gram-negative bacteria to grow. It also differentiates between gram-negative bacteria that ferment lactose and those that do not. This medium detects coliforms, which are gram-negative, non–spore-forming bacilli that ferment lactose to produce acid and gas in 48 hours at 35°C. Coliform organisms appear as purple colonies that have a green metallic sheen.

Procedure:

Various differential and selective media inoculated with different organisms will be provided. Study and record observations of these media using drawings and verbal descriptions. Your course instructor will provide specific explanations regarding these media.

LABORATORY EXERCISE

10 HEMOLYSIN PRODUCTION BY MICROORGANISMS

Materials Needed:

18- to 24-hour Blood Agar (BA) plate inoculated with *Streptococcus pyogenes*
18- to 24-hour Blood Agar (BA) plate inoculated with *Streptococcus mitis*
18- to 24-hour Blood Agar (BA) plate inoculated with *Enterococcus faecalis*
Blood Agar (BA) plate
Cotton swab (sterile)

Purpose:

To demonstrate the action of hemolysins on red blood cells (RBCs).

Background Information:

Hemolytic activity indicates pathogenicity in the *Streptococci*. Hemolysins are enzymes or toxins that destroy RBCs, releasing iron and other nutrients the bacterium uses for growth. The medium used for this experiment is blood agar, which consists of nutrient agar containing 5% whole sheep's blood. Organisms inoculated on blood agar demonstrate one of three different phenotypes, alpha- (α), beta- (β), or gamma- (γ) hemolysis. α-Hemolysins partially lyse RBCs, causing a greenish zone around bacterial colonies. β-Hemolysins completely lyse RBCs, leaving a clear zone around bacterial colonies. γ-Hemolysis is really no hemolysis at all, resulting in no clearing around bacterial colonies. Generally, members of the *Streptococci* that are β-hemolytic are the most virulent and are usually associated with acute fulminating infections. Hemolysin production is not confined to the *Streptococci*. Many other microorganisms may exhibit hemolytic activity.

Procedure:

A. *Demonstration of Red Blood Cell Hemolysis.*

Note the type of hemolysis demonstrated by *S. pyogenes*, *S. mitis,* and *E. faecalis* on previously streaked out blood agar plates. Record the results using drawings and verbal descriptions.

B. *Throat Swab using Blood Agar Plates.*

1. Using a sterile cotton swab, gently swab the back of the throat near the uvula/tonsilar area of a classmate.
2. Swab section one of a blood agar plate that has been divided into four quarters. Discard the swab in an appropriate container and then use a sterile loop to streak quadrants 2, 3, and 4 of the plate as described in Laboratory Exercise 5.
3. Incubate the plate at room temperature for 1 week.
4. Examine the plate for signs of hemolysis. Record the results using drawings and verbal descriptions.

LABORATORY EXERCISE

MOTILITY TEST MEDIUM

Materials Needed:

18- to 24-hour Nutrient Broth (NB) culture of *Proteus vulgaris*
18- to 24-hour Nutrient Broth (NB) culture of *Staphylococcus epidermidis*
Motility Test Medium (MTM) tubes (2)

Purpose:

To observe the mobility of some microorganisms.

Background Information:

Some microorganisms propel themselves through a liquid medium using fine hair-like appendages called flagella. These are very difficult to see unless stained by special procedures. Organisms with flagella are motile (able to swim through liquids). Flagella are classified by their positions on the bacterial cell. Polar flagella are located at one or both ends of a bacterial cell. Peritrichous flagella completely cover the surface of a bacterial cell.

Motility medium is the most commonly used medium for detecting motility in the clinical laboratory. Motile microbes introduced into this semi-solid medium migrate away from the point of inoculation and cause the medium to look cloudy. Nonmotile microbes will not migrate away from the point of inoculation and will form a distinct line of growth in the medium.

Procedure:

1. Inoculate a Motility Test Medium tube with one of the two strains using a needle (Fig. 11–1). Stab the needle into the center of the tube with a single even motion about one half of the depth of the medium. Using a second Motility Test Medium tube, repeat this inoculation step with the other bacterial strain.

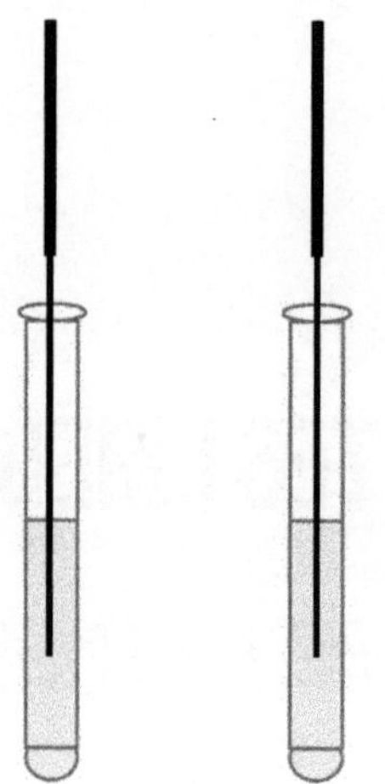

P. vulgaris *S. epidermidis*

Figure 11–1. Inoculation of Motility Test Medium. Stab the needle into the center of the tube with a single even motion about one half of the depth of the medium.

2. Incubate the tubes at room temperature for 1 week.
3. Record the results using drawings and verbal descriptions.

LABORATORY EXERCISE

12

CATALASE IN BACTERIA

Materials Needed:

18- to 24-hour Nutrient Broth (NB) culture of *Enterococcus faecalis*
18- to 24-hour Nutrient Broth (NB) culture of *Staphylococcus epidermidis*
Nutrient agar (NA) plates (2)
3% Hydrogen peroxide solution

Purpose:

To detect catalase activity in some microorganisms.

Background Information:

Aerobic microorganisms produce catalase. Catalase breaks down hydrogen peroxide (H_2O_2) to produce oxygen (O_2) and water (H_2O) and is not found in anaerobic or microaerophilic bacteria. The lack of catalase in these bacteria partially explains why oxygen is poisonous to them. When oxygen is available, bacteria produce hydrogen peroxide during their metabolism, but anaerobes cannot degrade it, whereas catalase producing bacteria can degrade it. Thus, the accumulation of hydrogen peroxide kills anaerobic and microaerophilic bacteria.

Purpose:

1. Streak out each strain, *E. faecalis* and *S. epidermidis*, for isolation on nutrient agar plates as in Laboratory Exercise 5.
2. Incubate the plates at room temperature for 1 week.
3. Add several drops of the 3% hydrogen peroxide solution to the bacteria on the plates. The presence of catalase in the bacteria is indicated by a trail of oxygen (O_2) bubbles arising from the colony.
4. Record the results using drawings and verbal descriptions.

LABORATORY EXERCISE 13

DIFFERENTIAL STAINING OF BACTERIA: THE ACID-FAST STAIN

Materials Needed:

18- to 24-hour Nutrient Agar (NA) plate of *Mycobacterium smegmatis*
18- to 24-hour Nutrient Agar (NA) culture of *Enterobacter aerogenes*
Glass slides

Purpose:

To identify *Mycobacteria* using the acid-fast staining technique.

Background Information:

The acid-fast stain is a differential stain. It identifies bacteria in the genus, *Mycobacterium*, which contains the microbes that cause tuberculosis (*Mycobacterium tuberculosis*) and leprosy (*Mycobacterium leprae*). A smear of the test microorganism is stained with a solution of carbolfuchsin (the primary stain) in the presence of heat. Acid-fast bacteria have fats and waxes in their cell membrane, making their cell wall impermeable to carbolfuchsin, unless they are heated. All bacteria will stain red with carbolfuchsin in the presence of heat. The smear is then decolorized with acid-alcohol. Non–acid-fast bacteria will lose the carbolfuchsin dye. Acid-fast bacteria retain the dye because the fats and waxes in their cell membrane, when cool, make their cell wall impermeable to acid-alcohol. Counterstaining the smear with methylene blue makes it possible to see acid-fast–negative bacteria. Thus, acid-fast–positive bacteria appear red, while acid-fast–negative bacteria appear blue under the oil immersion objective of a microscope. Because of its specificity, the acid-fast–staining technique is a primary tool for diagnosing tuberculosis and other infections caused by *Mycobacteria*.

Procedure:

1. Prepare a mixed smear of *M. smegmatis* and *E. aerogenes* as described in Laboratory Exercise 3 (Fig. 13–1). You will add the liquid culture of *E. aerogenes* to the slide first using a sterile loop. Sterilize the loop again and add a small sample of *M. smegmatis* from the plate to the slide. Mix these bacteria well on the slide because they tend to clump.

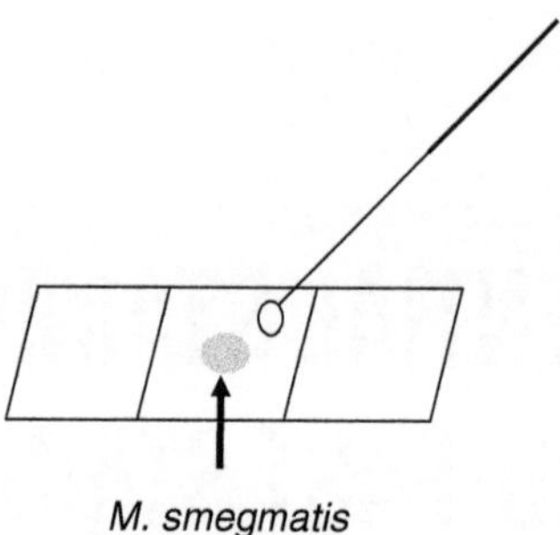

M. smegmatis
E. aerogenes

Figure 13–1. Mix *M. smegmatis* and *E. aerogenes* in a single smear.

2. Flood the smear with carbolfuchsin stain on a heating block at the lowest setting for 5 minutes. Heat is required for the dye to penetrate the acid-fast bacterial cell wall. Keep adding a drop of dye to the slide every 30 seconds to keep the dye from drying out.
3. Take the slide back to the sink and rinse the slide with water.
4. Holding the slide at a 45° angle, wash the smear with acid-alcohol until it runs colorless from the slide (about 5–10 seconds).
5. Rinse the slide with water.
6. Counterstain the smear by flooding it with methylene blue for 2 minutes.
7. Rinse the slide with water to wash off excess counterstain and blot it dry.
8. Examine the smear using the oil immersion objective of a microscope. Record the results using drawings and verbal descriptions.

LABORATORY
EXERCISE

14

CARBOHYDRATE FERMENTATION

Materials Needed:

18- to 24-hour Nutrient Broth (NB) culture of *Escherichia coli*
18- to 24-hour Nutrient Broth (NB) culture of *Alcaligenes faecalis*
18- to 24-hour Nutrient Broth (NB) culture of *Staphylococcus epidermidis*
Lactose fermentation liquid medium (3)

Purpose:

To identify bacteria that ferment lactose to produce acid and gas.

Background Information:

Some bacteria ferment lactose to produce acid and gas. The addition of the pH indicator, phenol red, detects the production of acid. Phenol red is red at a neutral pH and is yellow at an acidic pH. Bacteria that produce acid cause phenol red containing medium to turn yellow. Durham tubes that sit upside down in the culture tube detect gas production (Fig. 14–1). Bacteria that ferment lactose to produce acid and gas in lactose liquid medium cause the medium to turn yellow and produce an air bubble in the Durham tube. Some bacteria may produce acid without gas, but no bacterium produces gas without producing acid. If a bacterium does not ferment lactose, no color change occurs.

Procedure:

1. Inoculate each strain, *A. coli*, *A. faecalis,* and *S. epidermidis*, into a lactose tube (Fig. 14–1).

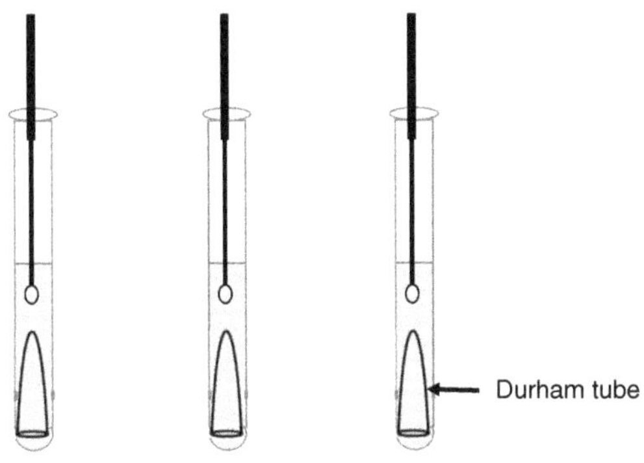

Figure 14–1. Lactose fermentation medium containing a Durham tube.

2. Incubate the cultures at room temperature for 1 week.
3. Record the color of the growth medium and the presence or absence of an air bubble in the Durham tube.

LABORATORY EXERCISE

15

TRIPLE SUGAR IRON AGAR

Materials Needed:

18- to 24-hour Nutrient Broth (NB) culture of *Escherichia coli*
18- to 24-hour Nutrient Broth (NB) culture of *Proteus mirabilis*
18- to 24-hour Nutrient Broth (NB) culture of *Staphylococcus epidermidis*
Triple Sugar Iron (TSI) agar slant (3)

Purpose:

To test for the fermentation of glucose, lactose, or sucrose and the production of hydrogen sulfide gas by some bacteria.

Background Information:

Triple Sugar Iron (TSI) agar medium tests for the fermentation of three sugars: glucose, lactose, and sucrose. It also detects for the production of carbon dioxide gas and hydrogen sulfide gas. This medium plays an important role in identifying many microorganisms, particularly enteric bacilli.

When TSI medium is made, it is allowed to solidify in a slanted position so that it has a slanted surface (Fig. 15–1A). The indicator in the medium is phenol red. Under alkaline conditions (K), the indicator appears red or orange but changes to yellow under acidic conditions (A). Lactose and sucrose fermentation results in a color change to yellow in the slant portion of the tube, while glucose fermentation results in a color change to yellow in the butt portion of the tube (Fig. 15–2). In addition, TSI medium contains iron, which reacts with hydrogen sulfide (H_2S) to produce a black color. If the bacterium produces hydrogen sulfide, a black precipitate will appear in the butt portion of the tube. Some bacteria may also produce CO_2 gas as exhibited by bubbles/cracks in the medium or even lifting of the slant.

Procedure:

1. Sterilize an inoculating needle using an incinerator and dip the needle in a broth culture containing one of the test microorganisms.
2. Inoculate a TSI agar slant by stabbing the needle through the slant to the butt (Fig. 15–1A).
3. Pull the needle out of the slant and streak the surface of the slant from the site of the stab (Fig. 15–1B).

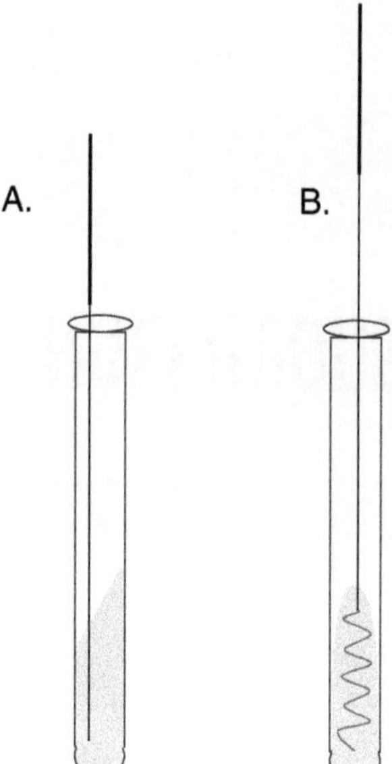

Figure 15–1. Inoculation of a triple sugar iron (TSI) slant. (A) Stab the needle to the butt of the slant. (B) After pulling the needle out of the agar, streak the surface of the slant.

4. Incubate the tube at room temperature for 1 week.
5. Report the results, slant/butt, as K/K, A/K, or A/A. K denotes no acid production and no fermentation, and A denotes acid production and fermentation (Fig. 15-2).

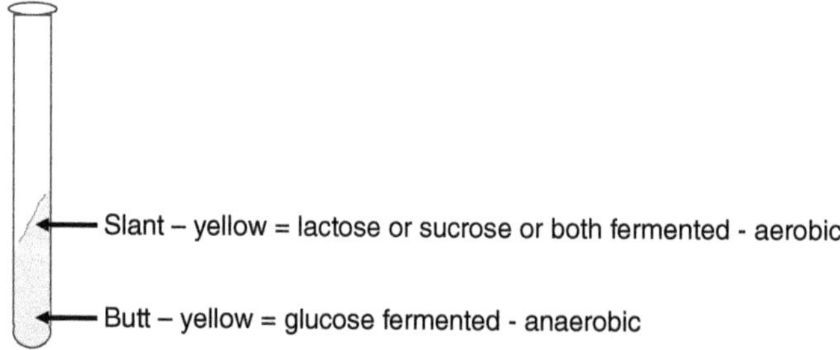

Figure 15–2. The slant and butt of a TSI slant. Red or orange indicates alkaline conditions (K) and no fermentation. Yellow indicates acidic conditions (A) and fermentation.

6. Report the results for hydrogen sulfide production. If the butt appears black, record the result as A/A + H_2S.
7. Finally, if cracks appear in the agar, report a positive result for carbon dioxide (CO_2) gas production.

LABORATORY EXERCISE

16 THE EFFECT OF DESICCATION ON MICROORGANISMS

Materials Needed:

Week 1
18- to 24-hour Nutrient Broth (NB) culture of *Bacillus subtilis*
18- to 24-hour Nutrient Broth (NB) culture of *Staphylococcus epidermidis*
Sterile test tube, capped (2)

Week 2
Transfer pipette, sterile (2)
Nutrient Broth (NB) (10 mL), sterile

Purpose:

To show that some microorganisms can withstand drying while others cannot.

Background Information:

All bacteria live as actively metabolizing and growing vegetative cells. Some bacteria undergo sporulation to produce endospores, resting cells which are resistant to heat, pH changes, chemicals, and drying. *Staphylococcus* forms an exoskeleton-like spore that also survives drying, and organisms in a biofilm are less sensitive to drying. Dried-out endospores will germinate and form vegetative cells when mixed with fresh growth medium. Dried-out *Staphylococcus* or biofilm bacteria will also resume growth when nutrients and water are restored. However, many non–endospore-forming bacteria only form vegetative cells that are sensitive to desiccation. Thus, many dried-out non–endospore-forming bacteria will not grow when mixed with fresh growth medium.

Procedure:

Week 1

1. Remove the cap from a sterile test tube.
2. Using a loop, spot a loopful of one of the test strains on the bottom of the test tube (Fig. 16–1). Place the cap back on the tube.

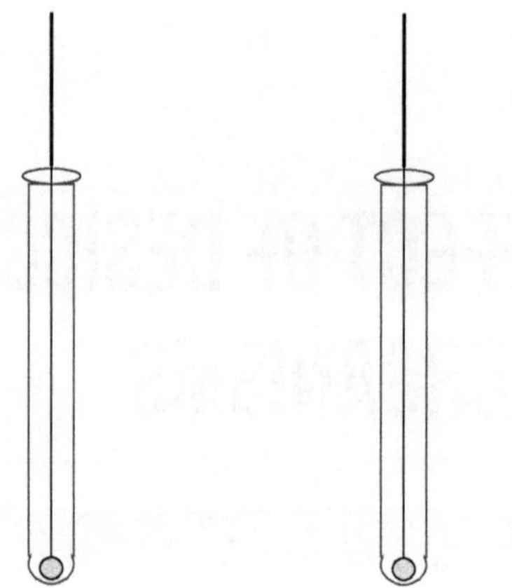

Figure 16–1. Desiccation susceptibility test. Place a loopful of each bacterial culture on the bottom of two sterile test tubes. Cap the tubes and allow the drops of bacteria to dry at 37°C for 1 week.

3. Repeat steps 1 and 2 for the other test strain.
4. Place the tubes in a 37°C incubator for 1 week to allow the drops to dry out.

Week 2

5. Remove the caps from the test tubes and use a sterile transfer pipette to aseptically add 2 mL of nutrient broth to each of the two test tubes. Place the caps back on the test tubes and incubate them at room temperature for 1 week.

Week 3

6. Turbidity (cloudiness) indicates growth, and no turbidity indicates no growth. Record growth or no growth for each bacterium.

LABORATORY EXERCISE

17 CHEMICAL DISINFECTION

Materials Needed:

18- to 24-hour Nutrient Broth (NB) culture of *Bacillus subtilis*
18- to 24-hour Nutrient Broth (NB) culture of *Staphylococcus epidermidis*
18- to 24-hour Nutrient Broth (NB) culture of *Alcaligenes faecalis*
Transfer pipettes
Test tube, sterile (1)
Nutrient Broth (NB), sterile tubes

Disinfectants to be tested:
- 2% Lysol solution
- 70% Ethanol solution
- Zephiran chloride solution
- 1:80 Phenol solution
- 1:100 Phenol solution

Purpose:

To show how some common disinfectants differ in controlling the growth of different bacteria.

Background Information:

A wide variety of chemical antimicrobial agents control the growth of microbes. Chemotherapeutic agents, such as antibiotics, are used to control infections in patients (Laboratory Exercise 21). Antiseptics are chemical agents that decrease the number of microorganisms on living tissue. Disinfectants are chemical agents that decrease the number of microorganisms on inanimate (nonliving) objects. Bactericidal agents kill bacteria, and bacteriostatic agents inhibit the growth of bacteria. Disinfectants and antiseptics may be bactericidal or bacteriostatic.

Many commercially available antimicrobial agents differ in their effectiveness. The phenol coefficient is one of the standards that is used to evaluate a chemical agent. Since phenol is a major component of many antimicrobials, the phenol coefficient directly compares the activity of the test chemical agent to phenol. The phenol coefficient only measures the efficacy of bactericidal compounds.

Procedure:

1. Using a transfer pipette, transfer 1 mL of disinfectant to a sterile test tube (Fig. 17–1). Each student will test one disinfectant and one bacterium. The disinfectant and bacterium will be assigned by the instructor. All groups will add data to a common table on week 2.
2. Using a transfer pipette, transfer 0.25 mL of test bacterium to the disinfectant.

3. After 1, 5, 10, and 20 minutes, use a sterile loop to transfer a loopful of the disinfectant/culture mixture to a tube of nutrient broth.

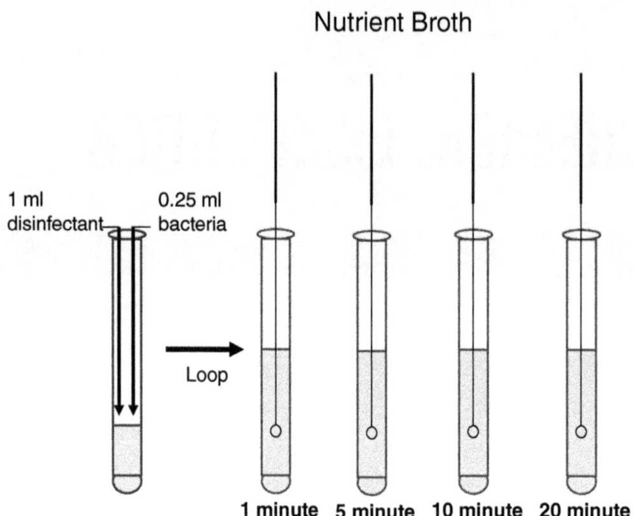

Figure 17–1. Disinfectant susceptibility test. Mix 1 mL of disinfectant with 0.25 mL of bacterial culture. At the indicated time period, transfer a loop of disinfectant/culture mixture to a fresh nutrient broth tube.

4. Discard the tube containing the disinfectant/culture mix. Incubate the inoculated nutrient broth tubes at room temperature for 1 week.
5. Record the results as growth (turbidity) or no growth (no turbidity) using Table 17–1. The amount of turbidity should be indicated by −, +, ++, and +++ to show the relative cloudiness of the tubes. These results should be added to the group table. After studying the class results, indicate how time exposure influences the effectiveness of the antimicrobial agents against each bacterial strain. In addition, note similarities and differences of the effectiveness of each antimicrobial agent against each bacterial strain.

Table 17–1. Growth of different bacterial strains after exposure to different antimicrobial agents.

	B. subtilis				*S. epidermidis*				*A. faecalis*			
Disinfectant	**1 min**	**5 min**	**10 min**	**20 min**	**1 min**	**5 min**	**10 min**	**20 min**	**1 min**	**5 min**	**10 min**	**20 min**

No Growth = -, Light Growth = +, Moderate Growth = ++,w Heavy Growth = +++.

LABORATORY EXERCISE

18 THE EFFECT OF DIFFERENT LIGHT SOURCES ON MICROORGANISMS

Materials Needed:

Nutrient Broth (NB) culture of *Bacillus subtilis*
Nutrient Agar (NA) plates (4)
Ultraviolet (UV) light source
High-intensity light source

Purpose:

To determine the effects of different light sources on the survival of bacteria.

Background Information:

The energy of light at wavelengths of 400 to 700 nm (high light = normal lighting) in the laboratory is too low to kill most bacteria. However, the energy of ultraviolet light at wavelengths of 100 to 400 nm is high enough to kill bacteria. It causes the formation of thymine dimers in DNA, blocking replication and transcription. Bacteria die because they cannot copy their chromosomes or make proteins. UV light does not pass through barriers such as glass or dirt. Thus, it does not sterilize objects well and is a much better disinfectant. It is especially useful for disinfecting hospital surgery rooms and microbiology laboratories.

Procedure:

1. Streak out four nutrient agar plates with *B. subtilis* and label them as shown in Figure 18–1.

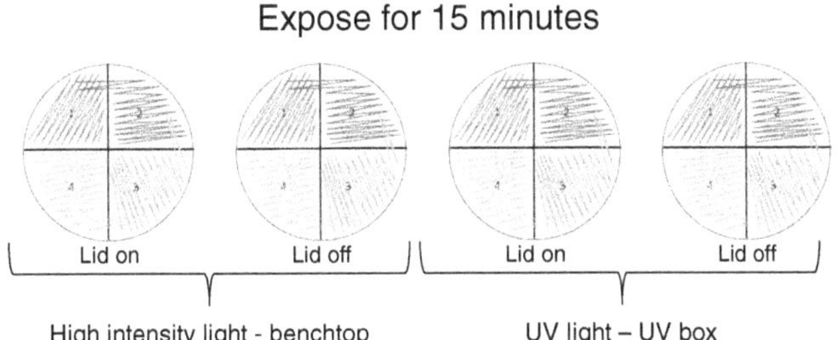

Figure 18–1. Exposure of *B. subtilis* to high-intensity and ultraviolet light. Streak out four nutrient agar plates of *B. subtilis* and expose them with the lid on or off to high-intensity and UV light for 15 minutes.

2. Expose two of the plates, one with the lid on and the other with the lid off, to the ultraviolet light for 15 minutes (the instructor will add the plates to the UV light box).
3. Expose the other two plates, one with the lid on and the other with the lid off, to classroom (high-intensity) light on the bench top for 15 minutes.
4. Incubate the plates at room temperature for 1 week.
5. Record the results using drawings and verbal descriptions.

LABORATORY EXERCISE

19 THE EFFECTS OF COLD TEMPERATURES ON MICROORGANISMS

Materials Needed:

Nutrient Broth (NB) culture of *Bacillus subtilis*
Nutrient Broth (NB) culture of *Staphylococcus epidermidis*
Nutrient Broth (NB) culture of *Escherichia coli*
Transfer pipettes, sterile
Nutrient Broth (NB), sterile tubes (3)

Purpose:

To determine the effect of cold on bacterial growth and survival.

Background Information:

Most bacteria that cause infections are mesophiles that grow best at body temperature, 37°C, because they reside in the human body. Their metabolic enzymes also function best at this temperature. Refrigeration (4°C) and freezing (0°C and below) retard enzymatic reactions in mesophiles and decrease their growth rate. When the temperature increases, bacterial enzymatic activity also increases, and the bacteria resume growing and multiplying. Thus, cold temperatures preserve foods and certain other materials but do not sterilize them.

Procedure:

1. Using a sterile transfer pipette, add one drop of test bacterium to three tubes of sterile nutrient broth (Fig. 19–1). Label each tube with the name of the bacterium and the incubation location, freezer, refrigerator, or room temperature. Each group will test one bacterial strain.

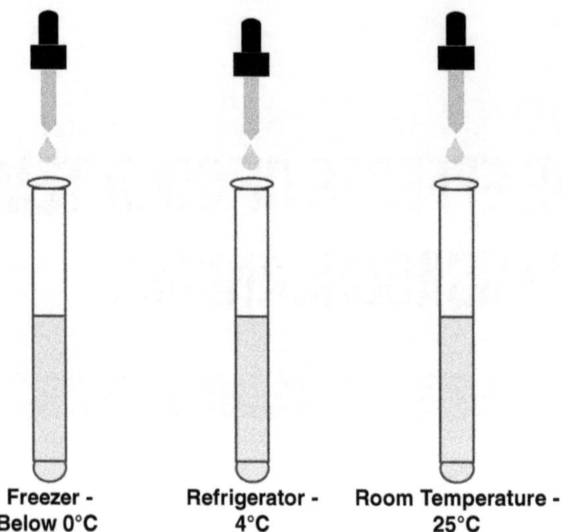

Figure 19–1. Each group will inoculate three nutrient broth tubes with one of the bacterial strains. Then, the first tube will be incubated in a freezer, the second in a refrigerator, and the third at room temperature.

2. The instructor will place racks on the counter for the collection of tubes to be placed in the refrigerator and freezer. Place the first tube in the freezer rack, the second tube in the refrigerator rack, and the third tube will be placed at room temperature for 1 week. Each bacterial strain will be incubated under all three conditions.
3. The instructor will remove the tubes from the refrigerator and freezer. After allowing the frozen culture to thaw, record the results as growth (turbidity) or no growth (no turbidity) for all tubes.
4. Place all the tubes at room temperature for another week, and record the results for each tube as growth (turbidity) or no growth (no turbidity). Determine if the temperature of incubation was bacteriostatic (prevented growth) or bactericidal (killed the organisms).

SKIN ANTISEPSIS

Materials Needed:

Nutrient Agar (NA) plates (3)
Alcohol wipes (2)

Purpose:

To determine the effects of an antiseptic (soap or isopropanol) on the presence of bacteria on the skin.

Background Information:

Although the skin is sterile during fetal development, bacteria colonize it when a baby enters the birth canal. While the normal biota consists of microbes that are nonpathogens, on occasion, the transient biota will harbor pathogens. Before surgical procedures, scrubbing with soap for 7 to 8 minutes removes most transient flora.

This experiment will examine the effectiveness of handwashing and antiseptic treatment on the skin biota. An antiseptic, such as soap or isopropanol, kills or removes pathogens on living tissue.

Procedure:

1. On the bottom side of three nutrient agar plates, use a marker to divide each into halves. Label the halves consecutively 1 through 6.
2. Open the lid. In half #1, touch the agar with several fingers and replace the lid (Fig. 20–1A).

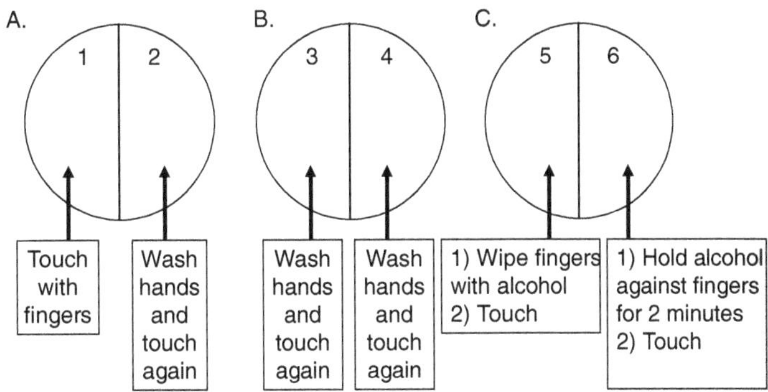

Figure 20–1. Procedure for skin antisepsis. A. Touch half #1 with your unwashed fingers. Wash your hands and touch half #2 with your fingers. B. Wash your hands a second time and touch half #3 with your fingers. Wash your hands a third time and touch half #4 with your fingers. C. Wipe your fingers with an alcohol wipe and touch half #5. Hold an alcohol wipe against your fingers for 2 minutes and touch half #6.

3. Wash your hands with soap and water and shake off the excess water. Open the lid and touch the agar in half #2 with your wet fingers. Replace the lid.
4. Wash your hands a second time and touch half #3 with your wet fingers (Fig. 20–1B).
5. Wash your hands a third time and touch half #4.
6. Wipe your fingers with an alcohol wipe and touch the agar in half #5 (Fig. 20–1C).
7. Hold an alcohol wipe against your fingers for 2 minutes and touch the agar in half #6.
8. Incubate the plates for 1 week at room temperature. Record your results using drawings and verbal descriptions.

LABORATORY EXERCISE

21 SENSITIVITY TESTING: THE KIRBY-BAUER METHOD

Materials Needed:

Tryptic Soy Broth (TSB) tube of *Alcaligenes faecalis*
Sensitivity disks (antibiotics and sulfonamides)
Nutrient agar plate (1)
Cotton swab, Alcohol wipe (1)

Purpose:

To determine the effects of different antibiotics on *A. faecalis*.

Background Information:

The dilution susceptibility test quantitatively measures the sensitivity of a bacterial strain to an antimicrobial drug. In this method, test tubes containing growth medium inoculated with the test strain are mixed with different concentrations of the antimicrobial drug. After incubating the mixtures, the tubes with growth (turbid) and without growth (clear) are recorded. The minimal inhibitory concentration (MIC) is the lowest concentration of the drug that prevents growth

The test tube dilution susceptibility technique has been replaced by a microtiter plate method. Microtiter plates contain multiple wells filled with medium inoculated with the test strain. Each well is then mixed with a different concentration of antimicrobial drug. Following 18 hours of incubation, the MIC is the lowest dilution of drug able to inhibit the growth (no turbidity) of the test bacterium.

The Kirby-Bauer method involves the diffusion of the antimicrobial drug into agar medium (Bauer, *et al.*, 1966; Petersdorf and Sherris, 1965). Bacteria are swabbed across the agar so that they cover the whole surface of a plate. Then, filter disks impregnated with an antimicrobial drug are placed on the surface of the agar. The agent diffuses through the agar, creating a concentration gradient with a high concentration near the disk and decreasing concentrations further away from the disk. The plates are then incubated to allow the bacteria to grow. Antimicrobial agents that inhibit the growth of or kill the bacterium create a zone of inhibition (clear area) around the disk, demonstrating that the bacterium is sensitive to the agent. Antimicrobial agents that do not inhibit growth will not create a zone of inhibition, demonstrating that the bacterium is resistant to the agent. Clinical experience established that the Kirby-Bauer test is a useful guide for selecting antimicrobial drugs to treat infectious diseases. This method is a qualitative test, yielding potential results of S (sensitive), I (intermediate), or R (resistant). Although the qualitative data from the Kirby-Bauer test are usually adequate for guiding the therapy of most infections, quantitative data from a dilution susceptibility test are desirable when drug dosage schedules must be monitored or when Kirby-Bauer test results are inapplicable, equivocal, or unreliable.

Procedure:

1. Dip a swab in a suspension of *A. faecalis*.
2. Swab a nutrient agar plate in three different directions so that the bacteria completely cover the whole surface of the agar (Fig. 21–1).

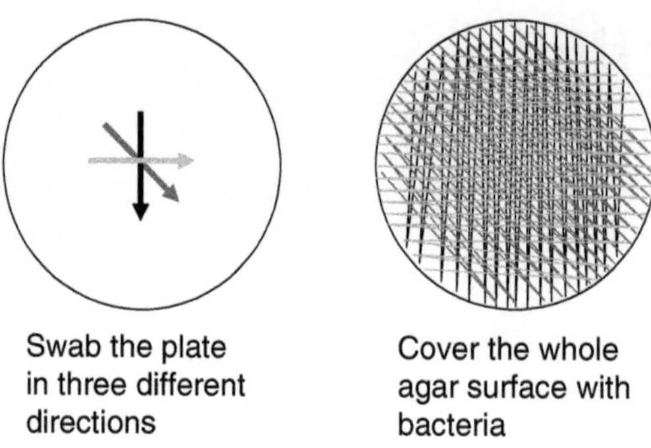

Figure 21–1. Swab a nutrient agar plate in three different directions so that the bacteria completely cover the surface of the agar.

3. Allow the plate to dry for 3 to 5 minutes.
4. Sanitize a pair of forceps with an alcohol wipe and use them to add antibiotic disks in an evenly spaced circle (Fig. 21–2).

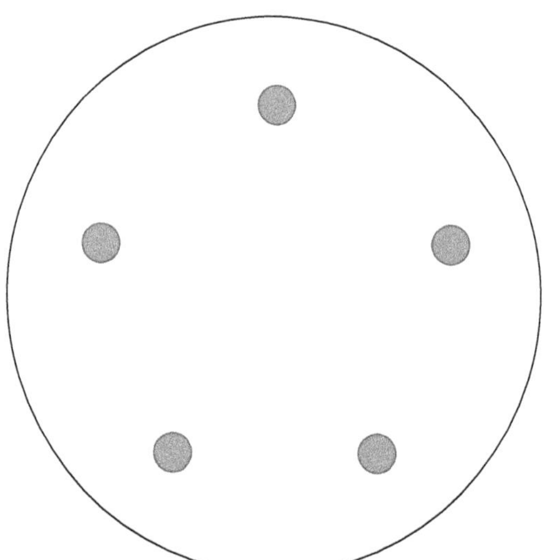

Figure 21–2. Place antibiotic disks in an evenly spaced circle on the plate.

5. Gently press the disks with forceps to ensure that they are flat on the plate.
6. Incubate the plate at room temperature for 1 week.
7. Measure the diameter in millimeters of the clear area formed around the disk (including the 6 mm disk). This clear area is the zone of inhibition. The size of the zone of inhibition indicates the effectiveness of the antibiotic.

LABORATORY EXERCISE

22 THE EFFECT OF OSMOTIC PRESSURE ON MICROORGANISMS

Materials Needed:

Nutrient Broth (NB) culture of *Bacillus subtilis*
Nutrient Broth (NB) culture of *Staphylococcus epidermidis*
Nutrient Broth (NB) tube
Nutrient Broth (NB) tube containing 5% sodium chloride
Nutrient Broth (NB) tube containing 10% sodium chloride
Nutrient Broth (NB) tube containing 25% sodium chloride

Purpose:

To determine the effects of salt concentration on the growth of bacteria.

Background Information:

A microorganism lives in an environment with constantly changing salt concentrations. The membranes of all living organisms are semipermeable. Water moves across them, but salt and sugars cannot. Osmosis is the movement of water across a membrane from a low solute (dissolved salts and sugars) concentration to a high solute concentration. In a hypotonic solution, the solute concentration surrounding a bacterium is very low in relation to the solute concentration within the bacterium (Fig. 22–1A). To equalize the solute concentration, water moves into the bacterium. The bacterial cells begin to swell, but the rigid cell wall prevents it from lysing (bursting).

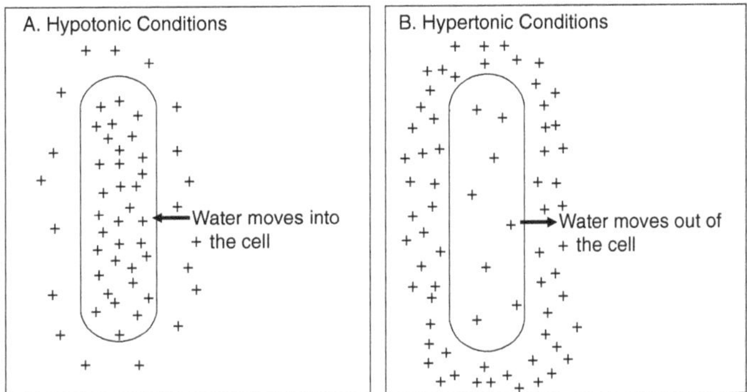

Figure 22–1. Osmotic conditions. (A) Hypotonic conditions—the solute concentration outside the cell is lower than that inside the cell. (B) Hypertonic conditions—the solute concentration outside the cell is higher than that inside the cell. The "+" symbol represents solute ions or molecules.

In an isotonic solution, the solute concentration outside the cell is equal to the solute concentration inside the cell. Water moves neither into nor out of the cell.

In a hypertonic solution, the solute concentration outside the cell is higher than the solute concentration inside the cell (Fig. 22–1B). Water moves out of the cell, and the cell membrane shrinks away from the cell wall. This condition has the same effect as desiccation (drying) because it prevents a bacterium from acquiring water. Some foods are prepared under hypertonic conditions to prevent spoilage. Salts preserve meats, and high sugar concentrations preserve jams. However, some microbes, such as molds, are more tolerant of hypertonic solutions than bacteria and will grow in this type of food. Thus, foods stored under hypertonic conditions are also sealed under anaerobic conditions because molds cannot grow in the absence of oxygen.

Procedure:

1. Use a loop to inoculate each salt broth medium (0%, 5%, 10%, and 25%) with one of the bacterial strains. Each group will test one of the bacterial strains.

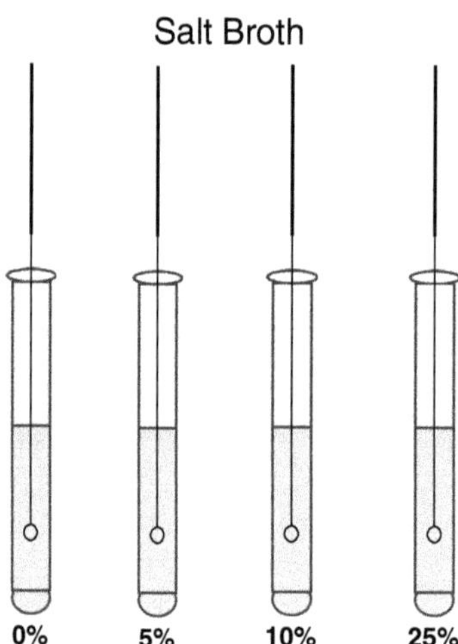

Figure 22–2. Inoculation of 0%, 5%, 10% and 25% salt broth.

2. Incubate the tubes at room temperature for 1 week.
3. Record the results as growth (turbidity) or no growth (no turbidity) using Table 22–1. At what salt concentration did you observe growth inhibition?

Table 22–1. Growth in salt broth medium.

	0% Salt	5% Salt	10% Salt	25% Salt
B. subtilis				
S. epidermidis				

No Growth = -, Light Growth = +, Medium Growth = ++, Heavy Growth = +++.

LABORATORY EXERCISE

EFFECTS OF PH ON MICROORGANISMS

Materials Needed:

Nutrient Broth (NB) culture of *Bacillus subtilis*
Nutrient Broth (NB) culture of *Staphylococcus epidermidis*
Nutrient Broth (NB) tube, pH 4
Nutrient Broth (NB) tube, pH 7
Nutrient Broth (NB) tube, pH 10

Purpose:

To determine the effects of pH on bacterial growth.

Background Information:

The pH indicates the hydrogen ion (H^+) concentration in a solution. It is calculated using the equation, $-\log([H^+])$, where $[H^+]$ is the hydrogen ion concentration. At pH 4, the solution is acidic and the H^+ concentration is 10^{-4} M. Acidophilic bacteria grow at acidic pH. At pH 7, the solution is neutral and the H^+ concentration is 10^{-7} M. Neutrophilic bacteria grow at neutral pH. At pH 10, the solution is basic (alkaline) and the H^+ concentration is 10^{-10} M. Alkalophilic bacteria grow at high pH. Since most bacteria are neutrophilic, slightly acidic (low pH) additives, such as acetic acid or citric acid, are used to preserve food. The high pH (alkaline) of many cleaning solutions also serves as a disinfectant.

Procedure:

1. Use a loop to inoculate each pH nutrient broth medium tube (pH 4, 7, and 10) with one of the test strains (Fig. 23–1). Each group will test one of the bacterial strains.

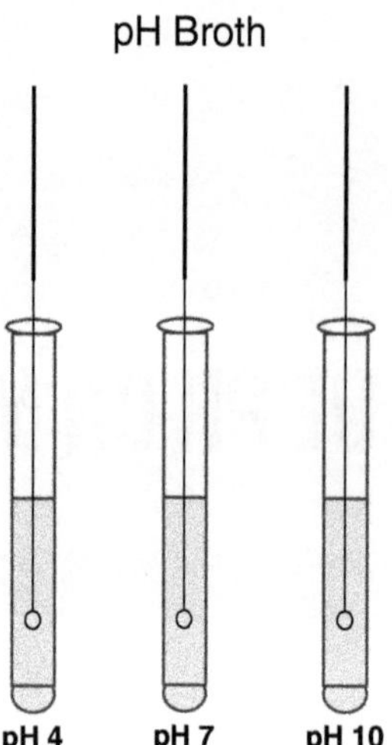

Figure 23–1. Inoculate each pH nutrient medium tube with a bacterial strain.

2. Incubate the tubes at room temperature for 1 week.
3. Record the results as growth (turbidity) or no growth (no turbidity) using Table 23–1. At what pH did you observe growth inhibition?

Table 23–1. Growth in pH nutrient medium

	pH 4	**pH 7**	**pH 10**
B. subtilis			
S. epidermidis			

No Growth = -, Light Growth = +, Medium Growth = ++, Heavy Growth = +++.

LABORATORY EXERCISE

24 MOIST HEAT STERILIZATION: BOILING AND AUTOCLAVING

Materials Needed:

Nutrient Broth (NB) culture of *Bacillus subtilis*
Nutrient Broth (NB) culture of *Staphylococcus epidermidis*
Nutrient Broth (NB) culture of *Escherichia coli*
Nutrient Broth (NB) tubes
1 mL pipette

Purpose:

To determine the effect of temperature on bacterial growth.

Background Information:

Dry heat cannot always be used for sterilization because it destroys some materials, such as liquids. Boiling, which heats solutions to 100°C at atmospheric pressure, kills vegetative cells but does not kill heat-resistant endospores. An autoclave heats objects at temperatures higher than 100°C. Increasing the pressure to 15 lb/in^2 allows it to elevate the temperature above 100°C to 121°C. Exposure to steam in an autoclave at this temperature for 15 minutes kills both vegetative cells and endospores. The autoclave, therefore, is one of the most valuable tools for sterilization in hospitals and laboratories.

Procedure:

1. Inoculate five nutrient broth tubes with a drop of one of the test strains (Fig. 24–1). Each group will test one of the bacterial strains.

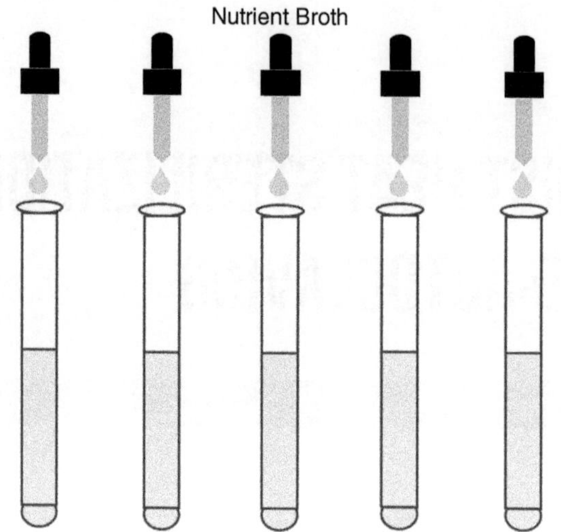

Figure 24–1. Inoculate five tubes of nutrient broth with a drop of one of the test strains. Autoclave one of the tubes. Expose the other four tubes to 100°C for 0, 1, 3, and 5 minutes.

2. Set one tube aside to be autoclaved (place in rack provided by instructor).
3. Leave one tube at room temperature and place the other three tubes in a boiling (100°C) water bath or a 100°C heating block.
4. Remove one of the tubes from the heat after 1 minute.
5. Remove another tube from the heat after 3 minutes.
6. Remove the last tube from the heat after 5 minutes.
7. Incubate all the five tubes at room temperature for 1 week (the instructor will place the autoclaved tubes out to incubate).
8. Record the results as growth (turbidity) or no growth (no turbidity) using Table 24-1. At what temperature did you observe growth inhibition?

Table 24–1. Growth after exposure to heat.

	Autoclaved	0 min	1 min	3 min	5 min
B. subtilis					
S. epidermidis					
E. coli					

No Growth = -, Light Growth = +, Medium Growth = ++, Heavy Growth = +++.

LABORATORY EXERCISE

25

BACTERIOLOGICAL TESTING OF WATER

Materials Needed:

Contaminated water
Uncontaminated water
Lauryl Sulfate Broth (LSB) tubes (2)
Brilliant Green Bile (BGB) tubes (1)
Eosin Methylene Blue (EMB) agar plate (1)
Nutrient Agar (NA) plate (1)

Purpose:

To detect fecal contamination in water.

Background Information:

Safe drinking water should not only be free of harmful chemicals but should also be free of pathogenic bacteria. Coliforms are Gram-negative, non–spore-forming rods, which ferment lactose to produce acid and CO_2 gas. Coliforms, which are also colon bacteria, are indicators of fecal contamination and for the possible presence of pathogens in water.

Lauryl sulfate broth (LSB) was one of the first tests used to detect coliforms in water. It is a selective and differential medium because it promotes coliform growth and detects lactose fermentation by the appearance of gas in a Durham tube. Growth with gas production is a positive result for coliforms in LSB.

The **Most Probable Number (MPN)** method uses lauryl sulfate broth (LSB) medium to estimate the number of coliforms in a water sample.

The MPN test involves placing three dilutions of water in five LSB tubes and incubating them for 1 week at room temperature. After recording the total number of positive tubes with growth and gas, a statistical table estimates the number of bacteria per milliliter of water.

This exercise will use not use the MPN test. Instead, it will use LSB, **b**rilliant **g**reen **b**ile (BGB) broth, and **e**osin **m**ethylene **b**lue (EMB) agar to detect coliforms in water. LSB will be inoculated with an unknown water sample and incubated for 1 week. If the result is positive (gas in the Durham tube), the bacteria in the LSB medium will be inoculated into BGB broth and streaked onto an EMB agar plate, and a nutrient agar plate.

BGB broth is a selective and differential medium. It is selective because brilliant green inhibits noncoliforms, and bile inhibits gram-positive bacteria. It is differential because when coliforms ferment lactose to produce acid and CO_2 gas, the brilliant green turns yellow, and gas forms air bubbles in a Durham tube. It is a better indicator of coliforms than LSB medium.

EMB agar is also a selective and differential medium for detecting coliforms. Eosin and methylene blue inhibit the growth of gram-positive bacteria and lactose-fermenting colonies appear purple with a green metallic sheen. Its properties are discussed in Laboratory Exercise 9.

Finally, a Gram-stain (see Lab 6) using the culture on the nutrient agar plate will verify that the isolated bacteria stain Gram-negative and are rod shaped.

Procedure:

1. Inoculate one lauryl sulfate broth (LSB) tube with 1 mL of contaminated water and a second LSB tube with 1 mL of uncontaminated water (Fig. 25–1).

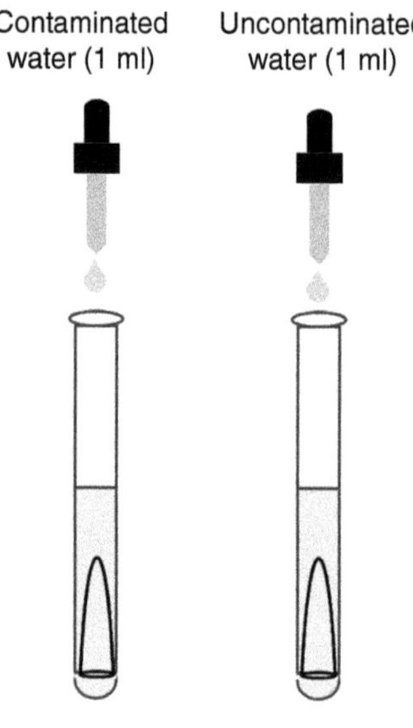

Figure 25–1. Inoculate two LSB broth tubes with 1 mL of water: one contaminated and the other uncontaminated.

2. Incubate the tubes at room temperature for 1 week.
3. Record your results using drawings and verbal descriptions. Turbidity (cloudiness) in the growth medium and an air bubble (gas) in the Durham tube indicate a positive result.
4. From a tube with a positive result, inoculate a tube of BGB medium, streak out a nutrient agar plate for isolation (Laboratory Exercise 5), and streak out an EMB agar plate (Fig. 25–2).

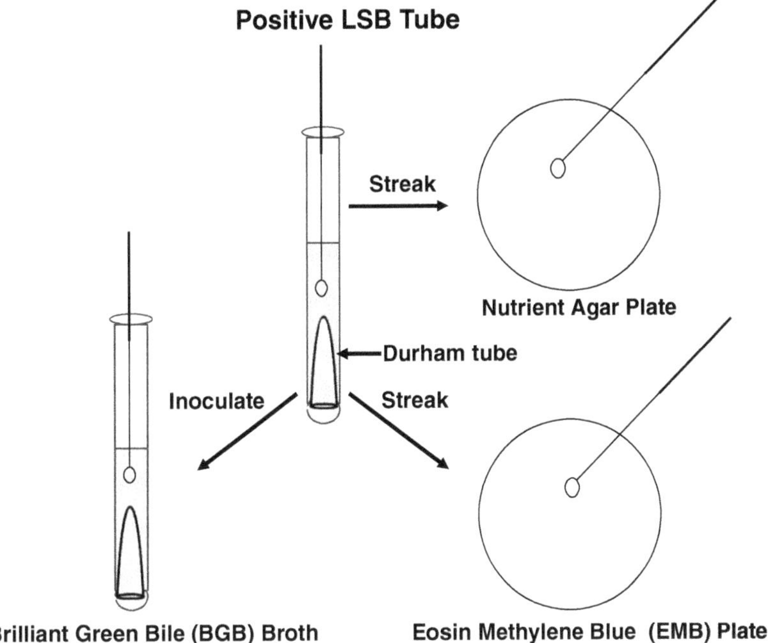

Figure 25–2. Confirmation of a positive result. Using a loop, inoculate bacteria from a positive result into BGB medium. Also, streak the bacteria onto a nutrient agar plate and onto an EMB plate.

5. Incubate the BGB culture and the two plates at room temperature for 1 week.
6. Perform a Gram-stain on bacteria that grew on the nutrient agar plate (Laboratory Exercise 6).
7. Record your results for the BGB medium, EMB plate, and Gram-stain using drawings and verbal descriptions.

LABORATORY EXERCISE 26

BACTERIA ON FRUITS AND VEGETABLES

Materials Needed:

Carrot or Apple
Nutrient Broth (NB) tube (2) Alcohol wipe (2)

Purpose:

To see if bacteria are present on the outside and inside of an apple or a carrot.

Background Information:

Bacteria are present on the surface of fruits and vegetables, but plants prevent microorganisms from penetrating the skin of fruits and vegetables. This exercise uses nutrient broth to detect the presence of bacteria on the surface and demonstrates that the inside of a fruit or vegetable is sterile if no damage has occurred to the skin or peel.

Procedure:

1. Wipe a knife and a pair of forceps with an alcohol wipe to sanitize them. Cut a small piece of peel from an apple or carrot and aseptically place it in a tube of nutrient broth (Fig. 26–1).

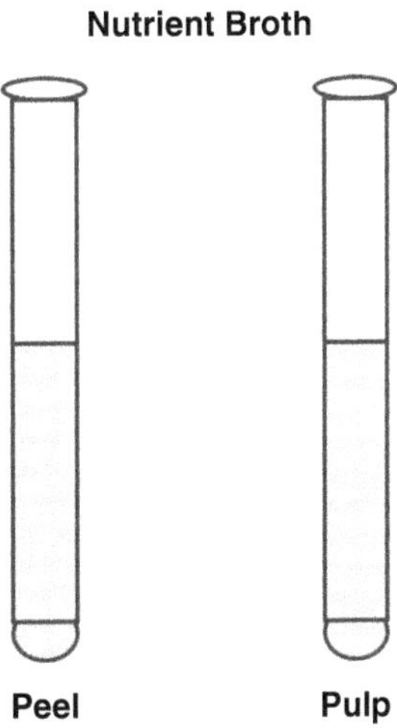

Figure 26–1. Inoculate two nutrient broth tubes with the peel or pulp of a carrot or apple.

2. Repeat step 1 taking an internal piece of the apple or carrot.
3. Incubate the tubes at room temperature for 1 week.
4. Record the results as growth (turbidity) or no growth (no turbidity).

LABORATORY EXERCISE 27

BACTERIA ON MEAT

Materials Needed:

Beef, Poultry, or Fish
Lauryl Sulfate Broth (LSB) tube (1)
Brilliant Green Bile (BGB) broth tube (1) Alcohol wipe

Purpose:

To see if coliforms are present in beef, poultry, or fish.

Background Information:

When animals are slaughtered to produce meat, fecal material usually contaminates the meat. This exercise uses lauryl sulfate broth and brilliant green bile broth to detect coliforms in beef, fish, or poultry. The results demonstrate the importance of cooking meat thoroughly to prevent disease.

Procedure:

1. Wipe a knife and a pair forceps with an alcohol wipe to sanitize them. Cut off two small pieces of beef, poultry, or fish.
2. Aseptically place one piece in a tube of LSB and the other in a tube of BGB broth (Fig. 27–1).

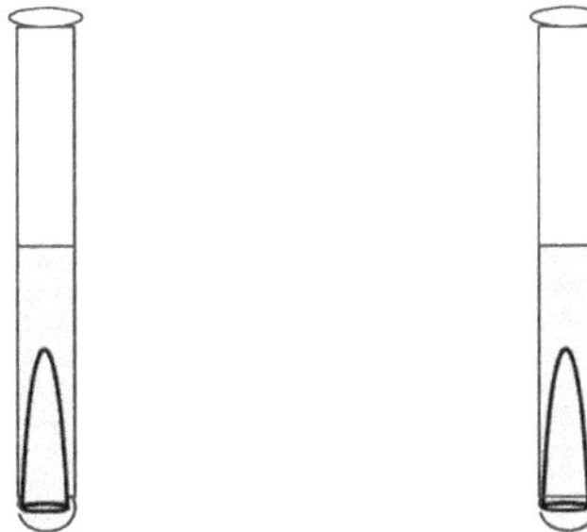

Figure 27–1. Using aseptic technique, place a piece of beef, poultry, or fish in an LSB broth and BGB tube.

3. Incubate the tubes at room temperature for 1 week.
4. Record the results for the LSB and BGB broth media using drawings and verbal descriptions. A coliform-positive result for LSB is growth (turbidity) and gas production (an air bubble in the Durham tube). A coliform-positive test for BGB broth is growth (turbidity), acid production (a color change to yellow), and gas production (an air bubble in the Durham tube).

LABORATORY EXERCISE 28

THE MAKING OF DOUGH

Materials Needed:

Petri dish (4)
Flour
Sugar
Yeast
Water

Purpose:

To show that yeast and sugar are required for bread dough to rise.

Background Information:

Yeast causes bread dough to rise by producing carbon dioxide (CO_2) gas. When mixed with flour, sugar, and water, it breaks down sugar, producing CO_2 and causing the dough to expand. This exercise shows that both yeast and sugar are required for bread dough to rise.

Procedure:

1. Put about 1/4 inch of flour in four Petri dishes.
2. Add additional ingredients as shown in Figure 28–1. When required, use a sprinkle of sugar and/or yeast. Be careful to not contaminate the spatulas with any ingredient other than the ingredient the spatula was found in.

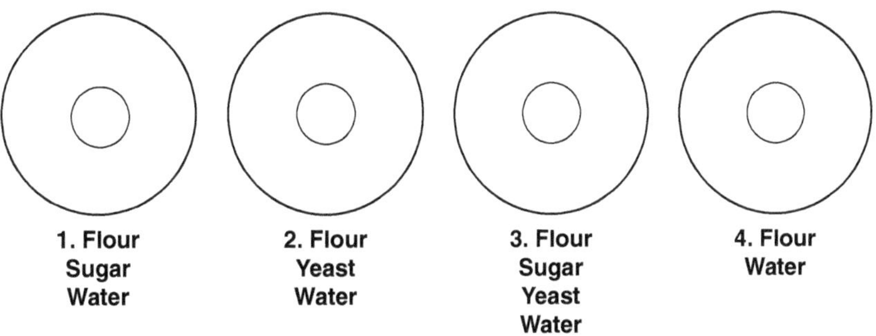

Figure 28–1. Required conditions for making bread dough.

3. Add enough water to make a paste. Push it into a ball at the center of the plate and circle the ball using a marker or wax pencil on the underside of the plate.
4. Put the lids on the plates and seal them with tape.
5. Incubate the plates at room temperature for 1 week.
6. Record the results for each condition using drawings and verbal descriptions.

LABORATORY EXERCISE

THE MAKING OF YOGURT CULTURES

Materials Needed:

Sterile, flat-bottom 50-mL tube
Yogurt starter culture
Milk
Sterile loop

Purpose:

To demonstrate the production of yogurt through the addition of probiotics to milk.

Background Information:

Probiotics are live bacteria added to foods to improve the health of the host. These probiotic organisms are bacteria commonly found living in the gut that provide many benefits to the host, including inhibition of pathogen growth, production of vitamins, reduction of cholesterol, improved gut health, and immunity. Yogurt is a good source of probiotics. During the production of yogurt, the microorganisms ferment sugars to produce lactic acid. Lactic acid changes the consistency of milk to a creamy texture and adds to the flavor of the yogurt. The bacteria commonly added to yogurt include *Lactobacillus acidophilus* (Gram-positive rods with variable shapes from short and plump to long and slender), *Lactobacillus bulgaricus* (long, filamentous, Gram-positive rods), *Streptococcus thermophiles* (Gram-positive cocci in chains), and *Bifidobacteria* (Gram-positive rods or branched bacteria). In this lab, you will produce yogurt using milk and bacteria found in a commercial preparation of yogurt.

Procedure:

1. Obtain a 50 ml conical plastic test tube, a small amount of starter yogurt culture, and milk.
2. Place 1 inch (15 mL) of milk into the plastic tube.
3. Using a sterile loop, place a small amount of yogurt into the milk and incubate at room temperature for 1 week (Fig. 29–1).

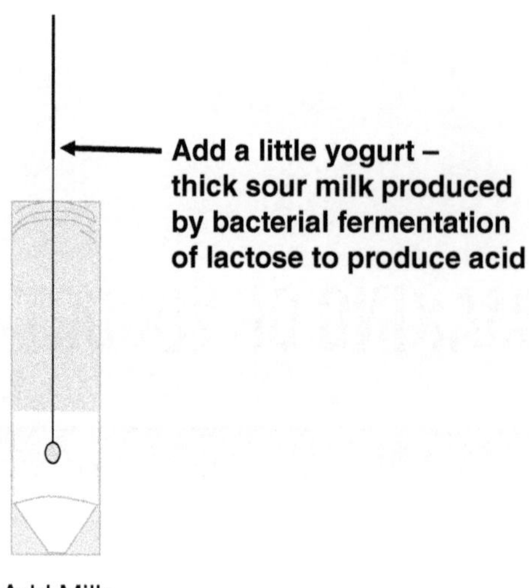

Figure 29–1. Inoculation of milk with yogurt.

4. Perform a Gram stain on the yogurt using a very small sample mixed into water on a slide. *Lactobacillus* or other probiotic organisms should be present. The organisms present will depend on the brand of yogurt used. Do not add too much yogurt to the slide, or the acidity in the yogurt will remove the crystal violet, giving the gram-positive organism the appearance of being gram negative.
5. After incubation, observe the quality (thickness and color) of the yogurt now present. It should be evident that acid produced by bacterial fermentation of the sugars in milk has caused a change in consistency of the milk to that expected for yogurt.
6. Repeat the Gram-stain on your yogurt and compare it to the Gram-stain you did on the commercial yogurt.
7. Describe the consistency of the milk, and draw and describe the organisms observed on the stains.

DENTAL CARIES

Materials Needed:

Snyder's agar tube
Petri dish (1)
Transfer pipette

Purpose:

To perform the Snyder test to demonstrate a person's susceptibility to dental caries (tooth decay).

Background Information:

Dental decay is caused by lactic acid and other organic acids dissolving the teeth. This acid production results from bacterial fermentation of simple sugars on the tooth surface. Sucrose (table sugar) is a disaccharide composed of the monosaccharides, glucose and fructose. Dental caries–causing organisms cleave sucrose into monosaccharides. In the process, fructose becomes available to be fermented to lactic acid. *Streptococcus mutans* is known to play a major role in tooth decay because it produces an exoenzyme, glucosyltransferase that forms large insoluble polymers of glucose called glucan. Dental plaque is glucan produced by caries-causing bacteria. The plaque contains living bacteria and adheres to the surface of the tooth, keeping the bacteria close to the tooth enamel. Lactic acid produced by bacterial fermentation of fructose beneath the plaque remains in contact with the tooth enamel. The enamel under the plaque is slowly decalcified and softened by lactic acid and decay begins. Although several microorganisms are involved in the process of acid production, only species in the genus *Lactobacillus* seem to be capable of lowering the pH enough to dissolve tooth enamel. The Snyder test measures dental caries susceptibility by detecting the presence of *Lactobacilli* in saliva.

Snyder test medium is designed to favor the growth of *Lactobacilli* and discourage the growth of most other species. This is accomplished by adjusting the pH of the medium to 4.8 and by adding glucose, a carbohydrate easily fermented by the bacteria. *Lactobacilli* thrive in the low pH and ferment glucose, producing more acid, which reduces the pH even more. The medium includes a pH indicator bromocresol green, which is green at pH 4.8 and above and yellow below pH 4.8.

The medium is autoclaved for sterilization, cooled to just over 40°C, and maintained in a warm water bath until needed. The still molten agar is then inoculated with saliva and allowed to grow for over 72 hours. In a normal Snyders test the agar tubes are checked at 24-hour intervals for any change in color. Yellow color indicates that fermentation has taken place and is a positive result. High susceptibility to dental caries is indicated if the medium turns yellow within 24 hours. Moderate susceptibility is indicated if the change is seen in 48 hours, and slight susceptibility if the change is seen in 72 hours. No change within 72 hours is considered a negative result.

Procedure:

1. Obtain Snyder medium from the water bath. Use a tube holder to transfer it to a cup for transfer to the bench.
2. Cool the agar until it is 55°C (warm enough to keep the agar melted, but cool enough to touch and prevent killing the bacteria).
3. Spit into a Petri dish (in the actual Snyder test, you would chew on a piece of wax to break up plaque in the mouth).
4. Transfer 1 mL of saliva to the Snyder media using a transfer pipette.

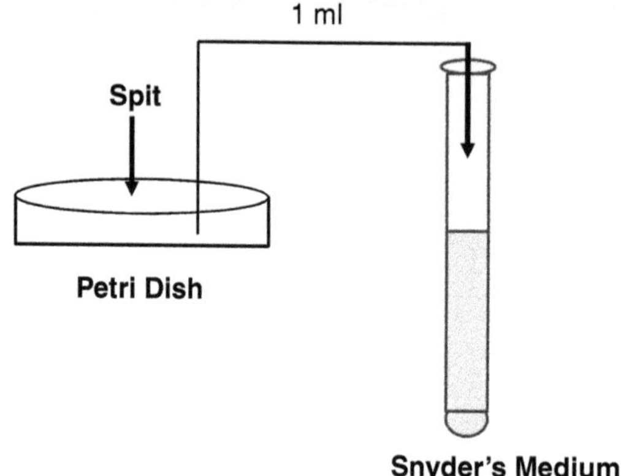

Figure 30–1. Inoculate Snyder media with saliva.

5. Mix the saliva in the media by rolling the tube between your hands.
6. Incubate the tube at room temperature for 1 week (in the actual Snyder test, you would incubate at 37°C and check it at 1, 2, and 3 days).
7. Observe tube and record results.

LABORATORY EXERCISE

31 STANDARD PLATE COUNT AND TOTAL COLIFORM COUNTS OF MILK

Materials Needed:

Raw milk for SPC and VRB
Pasteurized milk
Dilution blanks (99-mL bottle of saline and 9-mL tube of saline)
Standard Plate Count (SPC) agar bottle (molten)
Petri dishes, sterile (6)
Violet Red Bile (VRB) agar bottle (molten)
Transfer pipettes

Purpose:

To test milk for the number of total bacteria and for the number of coliforms to determine whether the milk is safe to drink.

Background Information:

Fecal contamination from the soil on cow udders frequently contaminates milk. To prevent the transmission of disease from contaminated milk, it must be pasteurized to heat-kill pathogens and undergo multiple tests before being certified safe for sale and consumption (Feng et al., 2002). Coliforms, Gram-negative, non–spore-forming rods that ferment lactose with the production of CO_2 gas, are organisms found in the colon. Because *Escherichia coli* is the coliform with the highest concentration in feces, it is used as an indicator of milk contamination. To verify the absence of *E. coli* in milk, it is tested using the Most Probable Number, BGB broth, EMB agar, Gram stain, the standard plate count (total number of bacteria), and the VRB agar technique. In this lab, you will learn the method for a standard plate count to determine the concentration of all bacteria in milk and the VRB agar method for determining the number of coliforms present.

Procedure:

This exercise is divided into two separate protocols, each described below in detail. Each group will perform the SPC test and the VRB test on one type of milk. The group will be paired with a group testing the second type of milk (raw or pasteurized). The groups will compare the results for both raw and pasteurized milk.

Standard Plate Count (total bacteria)

SPC agar is nutrient agar. All bacteria will grow on SPC agar, giving you a complete bacteria count.

1. Obtain one 99-mL and one 9-mL dilution blank, one SPC agar bottle (in a plastic cup), and three sterile Petri dishes. Label the three empty Petri dishes as follows: undiluted (1:1), 1:100, and 1:1,000. Label all plates with the type of milk you are testing (raw or pasteurized).
2. Using a sterile transfer pipette, place 1 mL of milk into the Petri dish labeled undiluted (Fig. 31–1).
3. Place 1 mL of milk into the 99-mL dilution blank and mix thoroughly. This is your 1:100 dilution.
4. From your 1:100 dilution, take 1 mL of sample and place it into your plate marked 1:100. This is your 1:100 concentration.
5. From your 1:100 dilution blank, take 1 mL of milk, place it into the 9-mL dilution blank and mix thoroughly. This is your 1:1,000 dilution.
6. Transfer 1 mL of the 1:1,000 dilution into the plate labeled 1:1,000. This is your 1:1,000 concentration.
7. Pour 25 mL of SPC agar into all Petri dishes (pour to cover bottom of plate) and swirl for even distribution. Allow to solidify.
8. Incubate the plates at room temperature for 1 week. Count the number of colonies on the best plate (the plate having around 50–100 colonies). Record the results.

REMEMBER: You must multiply by the dilution in order to correctly count the number of bacteria present per milliliter. For example, if there are 100 colonies on the plate labeled with the concentration of 1:1,000, then the actual count is 100 × 1,000 = 100,000 bacteria/mL.

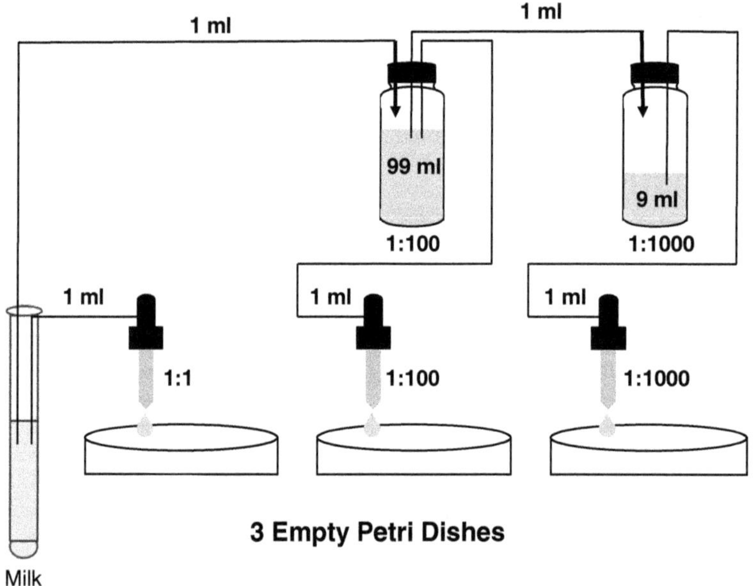

Figure 31–1. Method for making undiluted (1:1), 1:100, and 1:1,000 dilutions of milk. Add 1 mL of each dilution into the indicated plate. Then, pour approximately 25 mL of SPC agar into each plate.

Total Coliform Count

Violet Red Bile (VRB) agar is selective and differential. It is used to count the number of coliforms in milk. Coliforms are gram-negative, non–spore-forming rods that ferment lactose with the production of CO_2 gas. VRB agar is selective, as the bile in the agar inhibits the growth of gram-positive organisms. The Violet Red dye also inhibits noncoliform, gram-negative organisms. The agar is also differential, as lactose-fermenting organisms will produce dark red colonies.

1. Obtain one bottle of VRB agar and three sterile Petri dishes. Use the 1:100 milk dilution that was prepared for the SPC.
2. Label all three Petri dishes as follows: VRB 1:100 and the type of milk tested.
3. Using a transfer pipette, add 3.3 mL of the 1:100 dilution of milk to each plate (Fig. 31.2). The total amount of milk tested will be 10 mL (3.3 mL × 3 plates). Pour 25 mL of VRB agar into each of the three plates (cover the bottom of the dish). Swirl the plate for even distribution (until the milk is completely mixed into the agar). Allow the agar to solidify.
4. Incubate plates at room temperature for 1 week.
5. Counts of dark red colonies from all three plates.

REMEMBER: A 1:100 concentration was prepared. Therefore, you must multiply the total number of red colonies from all three plates by 100. This gives you the number of coliforms in 10 mL of milk. Record your results.

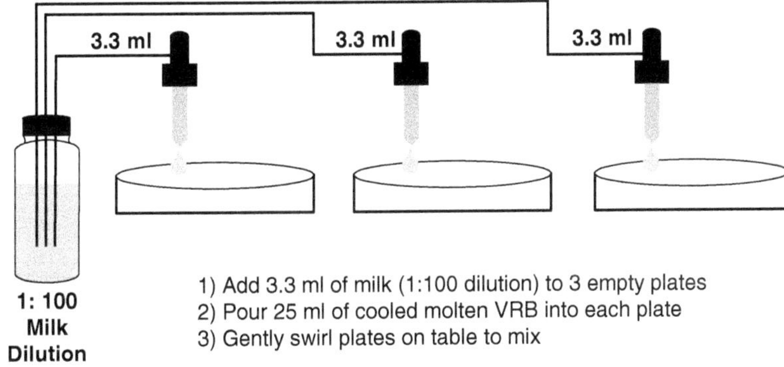

Figure 31–2. Total coliform count using VRB medium.

LABORATORY EXERCISE

PHOSPHATASE TEST

Materials Needed:

Raw milk for Phosphatase Test
Pasteurized milk
Phosphax
Glass test tubes, sterile (2)
CQG reagent
Transfer pipettes
n-Butyl alcohol

Purpose:

To test for the presence of alkaline phosphatase in milk to demonstrate if milk has been properly pasteurized.

Background Information:

This is the single most effective test of pasteurization in milk. Raw milk always contains the enzyme alkaline phosphatase, which is slightly more heat resistant than pathogens such as the tubercle bacillus. If pasteurization has been adequately performed, the enzyme will be destroyed. In the test, a small amount of milk sample is added to a solution of disodium dephenylphosphate (Phosphax) and heated. If active phosphatase is present, the compound splits, liberating phenol that will react with an indicator (CQG) to give a blue color. n-Butyl alcohol is added to extract the blue color from the milk. If phosphatase is present, the sample will turn blue. This indicates that the milk was not adequately pasteurized, as the enzyme is still active.

Procedure:

1. Obtain both a raw and pasteurized milk sample.
2. Pipette 1 inch of Phosphax solution into each of the two test tubes (Fig. 32–1).
3. Add 0.5 mL of the raw milk sample to one tube and 0.5 mL of pasteurized milk to the other. Label the tubes appropriately and mix the solutions by gently shaking the tubes (hold tube tightly and flick with your finger).

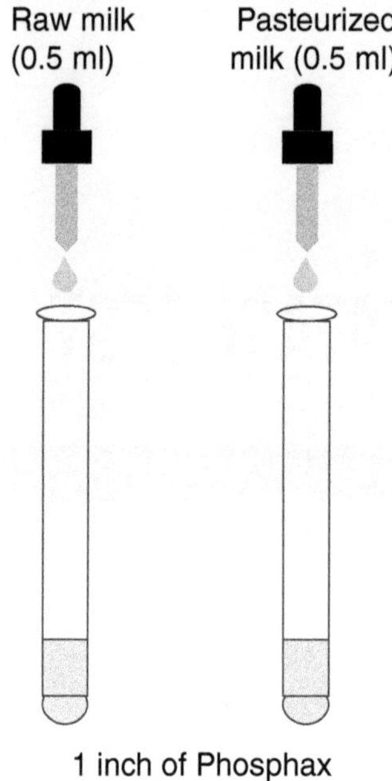

Figure 32–1. Phosphax test for alkaline phosphate enzyme in milk.

4. Place the tubes in a 37°C incubator for 20 minutes.
5. Remove the tubes from the incubator and add six drops of CQG reagent.
6. Mix the contents of the tubes (hold tube tightly and flick with your finger) and place them in a 37°C incubator for 5 minutes.
7. Remove the tubes and cool them under tap water briefly.
8. Add 3 mL of n-butyl alcohol to each tube.
9. Mix gently to extract the indophenol. A positive test for the presence of active phosphatase enzyme will yield a blue color. Record your results.

APPENDIX

ACCEPTED FORMAT FOR MICROBIOLOGY LABORATORY NOTEBOOKS

You should have a separate laboratory report for each experiment completed. The laboratory reports should be written in your own words and not be an exact copy of the laboratory manual. **Although you may have worked in groups, each individual should make their own observations and write their own laboratory reports without the assistance of other students.** Each page should be numbered and dated. Each laboratory report should include the following:

1. Title of Experiment
2. Purpose
   The purpose explains exactly why the experiment is being conducted and what the investigator is setting out to prove. This section may include any correlation that can be made between the experiment and a practical situation.
3. Materials
   The materials section lists the media, bacteria, and special supplies needed for the experiment.
4. Methods
   The methods are listed and numbered sequentially (not in paragraph form). They describe the experimental procedure in sufficient detail so that another investigator can repeat it and get the same results. Illustrations of techniques that explain the method should be included in this section.
5. Results
   The results tell the outcome of the experiment. This can be expressed in tables, graphs, drawings, or written form, depending on the type of data gathered. Examples are the shape of bacteria, color of bacteria or media, or cloudiness/turbidity of media.
6. Conclusions
   The conclusions state whether you accomplished the goals of your purpose. They explain what you learned and whether or not the results were expected. If the results were what you expected, explain why you expected those results. If the results were not what was anticipated, discuss possible reasons for the difference or changes that should be made to the method to correct any mistakes. Relate your conclusions to what you have learned in class or how it impacts your life. Example: 1. The color of an organism is a result, but what you learned from that color change (positive or negative for the test you were performing) is a conclusion; 2. If you studied bacteria from your body or from the environment, what did you learn about the bacteria that could impact you or others?

BIBLIOGRAPHY

1. Perry, L. A., and S. Senedak. *Food Microbiology Manual.* Youngstown, OH: Youngstown State University.
2. Perry, L. A., and P. D. Van Zandt. *Microbes in Focus: A Laboratory Manual of Microbiology for the Allied Health Professions.* Youngstown, OH: Youngstown State University.
3. Snyder, B. A. 1970. "Pitfalls in the Gram Stain." *Laboratory Medicine* 7, 41–44.
4. SharLab. *Handbook of Microbiological Media.* SharLab, S.L. Protocol. http://www.scharlab.com.
5. Petersdorf, R. G., and J. C. Sherris. 1965. "Methods and Significance of In Vitro Testing of Bacterial Sensitivity to Drugs." *The American Journal of Medicine* 39(5), 766–79.
6. Bauer, A. W., W. M. M. Kirby, J. C. Sherris, and M. Turck. 1966. "Antibiotic Susceptibility Testing by a Standardized Single Disk Method." *American Journal of Clinical Pathology* 45(4), 493–96.
7. Feng, P., S. D. Weagant, M. A. Grant, and W. Burkhardt. 2002. *BAM 4: Enumeration of* Escherichia coli *and the Coliform Bacteria.* U.S. Food and Drug Administration. *https://www.fda.gov/food/laboratory-methods-food/bam-4-enumeration-escherichia-coli-and-coliform-bacteria*

www.ingramcontent.com/pod-product-compliance
Ingram Content Group UK Ltd.
Pitfield, Milton Keynes, MK11 3LW, UK
UKHW051137260726
13967UKWH00010B/3095

9 781524 986216